AUTOMOTIVE FUEL INJECTION SYSTEMS
A TECHNICAL GUIDE

JAN P. NORBYE

Motorbooks International
Publishers & Wholesalers Inc
Osceola, Wisconsin 54020, USA

© Jan P. Norbye, MCMLXXXI
ISBN: 0-87938-139-6
Library of Congress Number: 81-16833

Printed and bound in the United States of America
Book and cover design by William F. Kosfeld
Cover photos courtesy of Robert Bosch Corporation

Motorbooks International is a certified trademark, registered with the United States Patent Office.

1 2 3 4 5 6 7 8 9 10

Motorbooks International books are also available in bulk quantity for industrial or sales-promotional use. For details write to Marketing Manager, Motorbooks International, P.O. Box 2, Osceola, Wisconsin 54020.

Library of Congress Cataloging in Publication Data

Norbye, Jan P.
    Automotive fuel injection systems.

    Includes index.
      1. Automobiles—Motors—Fuel injection systems.
I. Title.
TL214.F78N67   629.2′53   81-16833
ISBN 0-87938-139-6 (pbk.)   AACR2

Preface

Since becoming a full-time journalist twenty years ago, I have reported extensively on the subject of fuel injection. The principle bears promise of technical superiority, and the early examples of fuel-injected cars were glamorous high-performance vehicles, as if to corroborate the impression of advanced technology.

My first experience with fuel injection occurred in 1954, when Rudolf Oeser, then regional export manager of Daimler-Benz, took a 300 SL to Stockholm for a special exhibit. Sweden's economy was then very strong, leading the company to conclude that there would be a market for that type of car. Support for this conclusion came from the fact that the country was free of speed limits outside of built-up areas.

Before returning to Stuttgart, Rudolf Oeser took a side trip to Oslo, capital of currency-restricted, speed-limit-ridden (43.5 mph on the open highway!) Norway, where I was working on the monthly magazine of the leading automobile club.

We did not have good weather, and the demonstration was all the more convincing, as we rushed through a blanket of snow flurries whose accumulation on the cold roadway did not make the ride seem less hazardous, though it was conducted in perfect safety. My colleagues and I took turns riding in the passenger's seat with Oeser's expert demonstration driver at the wheel, a man who was making the most out of every short bit of straight road we came upon, to the extent of twice being able to get into fourth gear, with joyful disregard for all speed limits. After looking at the engine, and seeing the ram pipes and the elegant fuel lines to the injectors, I came away thoroughly convinced of the merits of fuel injection.

Three months later, I was at Le Mans where I watched Froilan Gonzalez and Maurice Trintignant bring home their 4.9-liter Ferrari V-12 in first place. Their car was equipped with triple 46 DCF/3 Weber carburetors, and my feelings about fuel injection were brutally rocked back into proper perspective.

In the years that followed, I was able to see how the adoption of fuel injection for both racing cars and production cars was rejected or delayed by Weber carburetors. Weber became in my eyes the worst enemy of fuel injection, surpassing the primitive applications of a superior principle in performance and efficiency by the sheer perfection of ancient technology!

Watching Fangio win the French Grand Prix at Reims in July 1954, supported by Kling and Herrmann, with their Mercedes-Benz W-196 'Silver Arrows,' did provide me with a boost for the belief that fuel injection was bound to triumph in the long run.

In December 1961, I had a chance to ride in the 300 SLR (powered by a three-liter version of the M-196 fuel-injected straight-eight) as Rudolf Uhlenhaut, chief engineer of car development, toured around the banked tri-oval Daytona Speedway at speeds up to 145 mph.

By mid-1963 I had racked up impressive mileages with the fuel-injected Peugeot 404 convertible, the Mercedes-Benz 220 SE, the new 230 SL and the Maserati 3500 Iniezione, each one stating a convincing case for fuel injection, each in its own way. But a few years later, Weber snatched back its previous monopoly on fuel-mixture-preparation hardware for Maserati engines, much to Lucas's dismay.

The switch to fuel injection was not a one-way street, as many had expected. Aston Martin and Triumph were other makes that reverted to carburetors after brief periods of producing fuel-injected models. These happenings were highly influential in the maturing of my own philosophy about fuel injection. Much as my natural enthusiasm leaned in favor of fuel injection, it was tempered by my admiration for a new and better carburetor.

The advent of electronic fuel injection in 1965 stirred things up in my mind, and new experiences soon came my way, including 6,000 miles in three weeks with a Citroen DS-23 Injection, a streamlined four-door sedan that would cruise at 115 mph for hours on end, with nothing more than a 140-cubic-inch four-cylinder engine under the hood!

There were also the Mercedes-Benz 350 SL and 450 SEL, the Alfetta from Alfa Romeo, Porsche 911, Peugeot 504 convertible, Saab 99 GLE and a host of others, either furthering the electronic revolution or exploring simpler, more conventional means in the form of Bosch's K-Jetronic.

For years, the trend was clear and strong, but after 1975, some flexing has set in, due to the need for lower-cost systems. This quest has produced impressive new solutions in the form of single-point injection systems and computer-controlled carburetors. And that brings us up to date.

I have been given invaluable help by a great many individuals and organizations in compiling this book.

Robert R. Hoge and Daniel D. Barnard of Bendix were tireless in their efforts to satisfy my demands, and Joe Brancik of the Graphic Arts Department gave a yeoman performance on the Bendix illustrations.

I am also greatly indebted to Frank Ulrich Breitsprecher and Eckhard D. Noelte of Robert Bosch AG, Peter J. Viererbl and Bernd Harling of Daimler-Benz AG, Michael Schimpke and Dirk Henning Strassl of BMW.

G. B. Jenkins of Lucas Industries, M.J.D. Hurn of Associated Engineering Ltd., Rag. Luigi Tura of Magneti Marelli, Robert H. Harnar and Charles L. Gumushian of the Ford Motor Company, Günter Böcker and M. Schantzli of Pierburg, Patricia Montgomery of Cadillac Motor Car Division and Jim J. Williams, public relations director of Chevrolet Motor Division, deserve everlasting gratitude for their generous assistance.

Jan P. Norbye
Les Issambres, November 1981.

Table of Contents

Introduction

Priorities in automotive engine design have undergone two major changes in the past twenty years. The emphasis used to be on maximum horsepower, consistent with the basic requirements of reliability and long life. Then came the Clean Air Acts of 1966 and 1970, which made an absolute priority of antipollution measures. After 1973 and the rapid quintupling of crude oil prices, priorities were redirected toward fuel conservation, which puts a premium on thermal and mechanical efficiency in an engine.

Horsepower, cleaner exhaust, lower fuel consumption—that just about sums up what we are all looking for in our next car. And we can get it, we are told, from one single thing: fuel injection.

Many engineers now hold the opinion that fuel injection will be a vital means of making further progress in operating efficiency without abandoning the gains that have been made in the emission control area. Others refuse to give up the carburetor, and hint that fuel injection is merely a passing phenomenon, a fad.

It seems disrespectful to talk of fads in automotive engineering, but it is a fact that the industry is not immune to the vagaries of fashion—examples abound. Now it's front-wheel drive. Suddenly, everybody in Detroit wants to make front-wheel-drive cars, after stubbornly resisting the European trend for generations. Before that, it was disc brakes and radial tires. Next, it may be fuel injection.

Call it a wave, to get away from the insulting term fad. The mass movement toward technical innovation is partly explained by cost considerations. The desire for engineering progress has always been tempered, in mass production, by the specter of higher costs.

Of course, all that's new isn't necessarily better. The way for the broad adoption of disc brakes was paved by the pressing need for better brakes. Making substantial improvements in drum-brake systems would cost a lot of money and alter the balance in the comparison with disc brakes, in favor of the latter. And as engineers know full well, after some experience with an initially expensive device they invariably find ways to make substantial cost cuts, either by simplifying the product or by developing new production methods; or a combination of both.

Rarely does a new feature show a total and indisputable superiority. It is possible to make very good drum brakes or very good bias-ply tires, for instance.

What happens, I think, is something like this: Coupled with the change in the industrial supply situation that occurs when a major car manufacturer undertakes a wide-scale change in the product, the publicity and marketing arms of the producers go into action to persuade the public that what they are doing is, in fact, progress.

It's easy to sell a superior principle, such as disc brakes or radial tires, and customers readily pay the higher price, in return for the advantages obtained.

Fuel injection has often been hailed as such a superior principle, and abundant evidence exists that car buyers were and are willing to accept the higher price. But when you closely examine the functions of a fuel-injection system, and those of a carburetor, this superiority is not so clear-cut. It could be questioned, or even challenged.

The differences lack contrast. The carburetor mixes fuel with air to form a combustible mixture. So does a fuel-injection system.

Are there really different principles at work? If you go deep enough into the details, yes. In the carburetor, fuel is sucked into the air, while fuel-injection squirts it into the air. It comes down to a choice between suction and pressure-spraying. The distinction is fine. If you have a system that works with very low-pressure injection, who can say that suction forces play no part in getting the fuel out of the nozzles?

But what about all those other claims enlisted in support of proving the superiority of fuel injection? Highly precise fuel metering, you have been told. That depends entirely on the type of measurement control that's built into the system. Accurate timing of fuel delivery? Not all systems have timed injection, but spray continuously.

Modern carburetors have been brought to a high state of perfection, and there is no reason to expect that it has reached its final stage.

Have the engineers let themselves be hypnotized by the vague promises of fuel injection? No, that's not the answer. I think it lies elsewhere. Disagreement among engineers as to the merits of fuel injection probably stems from differences in their perception of the qualities and shortcomings of the carburetor.

So many compromises are made in the carburetor that none of its functions can be performed optimally.

Fuel injection looks attractive immediately, on the strength of the realization that doing away with the carburetor eliminates its compromises and opens the way for more efficient engine operation. Getting the full benefits of fuel injection depends on designing and developing the engine to take full advantage of its promises. Take the question of power gains, for instance.

Why does fuel injection enable us to take more power out of an engine without increasing fuel consumption? Because fuel-injected engines will run smoothly with far higher compression than most carburetor engines can.

If automobile engines ran at a constant speed, the carburetor would not have a difficult task. As you know, they have a wide speed range, and must operate under variable load.

According to my friend Harry Mundy, Jaguar's engine designer, "The basic disadvantage of the carburetor is that there must be a constriction at the venturi, whose function it is to increase the velocity of the incoming air, thereby creating a vacuum to draw the fuel through the jets from the float chamber. Such a restriction limits the amount of mixture passing into the cylinders, and results in the power falling off at higher engine speeds. This can be overcome by fitting larger carburetors, but then the difficulties are transferred to the lower end of the speed scale, and flexibility is lost."

Intake manifolds are not, and perhaps cannot be, designed to facilitate things for the carburetor. The runners twist and turn, for their shape is determined mainly by the space available for them.

After leaving the carburetor, the mixture has to travel a sinuous route, with changes in flow velocity and pattern, different for each port. Because of these differences, it is not possible to obtain uniform mixture distribution, and some cylinders run with a richer mixture than others.

That does not give the complete picture, however, for the carburetor must be made to overcome three other fundamental problems of fuel-mixture preparation and engine operating conditions.

One difficulty is cold-starting. Another is the need for transient enrichment during acceleration. The third would not exist if cars traveled only in a straight line. However, roads have curves and streets have corners, and any change in the car's direction sets up a centrifugal force that affects all parts of the car and everything carried in it, including the fuel in the float bowl.

Fast cornering tends to force the fuel in the float bowl to climb up the wall, thereby raising the float to block further delivery. Result: fuel starvation. The driver experiences a sudden loss of power just when it is most wanted. The problem can be prevented by careful attention to float action at the design stage, which does not necessarily add to the cost of the carburetor. But the first and second difficulties require additional mechanisms to be added.

Carburetors must be designed to provide a richer mixture for starting. With cold air in a cold manifold, atomization is poor. A portion of the fuel separates from the air and clings to the walls of the manifold. The carburetor must make sure that enough fuel goes into the engine to have a mixture that can be ignited in the cylinders.

To do this, it has a choke mechanism, which can be operated manually or automatically. But in neither case can it be assumed that it responds with any accuracy to the operating conditions.

Then there's the problem of speeding up. Instantaneous opening of the throttle valve tends to lean-out the mixture because the speed of the fuel flow doesn't increase as quickly as air velocity. To avoid bucking or snatching, stumbling or hesitating, during sudden acceleration, an accelerator pump feeds additional fuel into the venturi. The accelerator pump, built into the carburetor, sprays a jet of fuel into the onrushing air at the critical moment. This pump is usually a simple plunger pump connected to the throttle linkage. The fuel from it is metered into the airstream in a steady discharge of short duration.

With fuel injection, the need for an acceleration pump, a choke mechanism and special means to assure continued fuel flow under high-lateral-force cornering conditions can be included as integral parts of the system and made to function automatically.

We must guard against thinking of fuel injection in terms of a rigid system and specific hardware. Instead, we must look at fuel injection as an operating principle that can be applied to several types of different systems. New variations are continually appearing, and the ultimate form of fuel injection may not yet exist.

Even for the time being, the fuel-injection engineer has many difficult choices to make. Where to inject? and When to inject? are the two primordial questions he must answer. He could inject (a) inside each cylinder; (b) into each port; or (c) farther upstream in the intake manifold.

The location of the injector nozzle dictates the timing to some extent. With direct injection (inside each cylinder), injection must occur in a short period during the compression stroke.

With port injection, timing is not critical. If it is timed, it should coincide with the valve opening period. Upstream injection is not sensitive to timing, but involves shortcomings in the control of mixture strength to the individual cylinders.

With timed port injection, the duration is partly a function of the quantity to be injected. With high-speed engines, a physical problem arises in that extremely minute quantities of fuel must be injected (and given time for atomization) in periods that can be measured only in milliseconds.

Direct injection requires a high-pressure pump which, aside from being noisy, consumes energy to pressurize the fuel. For production cars, the engineers' attention has always been focused on other systems, working with lower-pressure pumps consuming less energy. The lowest fuel-delivery pressures are obtained with electronic fuel-injection systems. In some companies, borderline systems that borrow from both carburetors and fuel injection, controlled by electronic means, are under investigation. In these companies, a new term has come into usage: electronic fuel management.

That may be the wave of the future. In the following text, we will examine the principles of mixture preparation, the combustion process and past and present fuel-injection systems, so as to give the reader a solid knowledge of the whole subject, from the basics to the frontiers of technology.

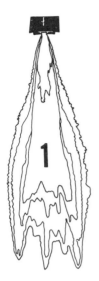

Injection or Carburetion?

As demonstrated in the introduction, the main attractions of fuel injection are to be found in the shortcomings of the carburetor. This fact makes the carburetor a logical starting point for any comprehensive examination of fuel-injection principles and hardware, and necessitates a detailed look at its construction and various functions.

Since the carburetor is, in a sense, the guard at the gate, letting air in or preventing it from entering, it's easy to see that the carburetor is the first link in controlling how the engine breathes.

During the intake stroke the piston should theoretically draw into the cylinder a volume of mixture at atmospheric pressure equal to the cylinder displacement. In practice the quantity of gas actually drawn in is always less than the theoretical amount.

The ratio between the theoretical amount and the actual amount is called volumetric efficiency. The usual value for a modern engine is eighty to eighty-five percent at full throttle. Poor volumetric efficiency does not necessarily hurt fuel economy but results in low power output relative to engine size. The principal reasons for failing to reach one hundred percent are:

1. Restrictions in the carburetor, and bends in the intake manifold and porting system, limiting gas flow to the cylinders.
2. Heating of the incoming charge by a hot intake port, or by other hot parts in proximity to the intake manifold, causing the air-fuel mixture to expand before entering the cylinders.
3. Hot exhaust gases remaining trapped inside the cylinder after the exhaust stroke.

The basic idea of the carburetor is to have an air passage with automatic fuel feed, self-regulated to suit the mass airflow as measured through a venturi. In principle, a carburetor consists of a fuel reservoir or float chamber and a fuel jet leading to the main venturi, or air passage, equipped with a throttle valve. A constant level is maintained in the reservoir by a float that opens or closes the fuel valve through a simple linkage. Fuel supply is assured by the action of the fuel pump, bringing raw gasoline from the tank.

The carburetor works by mixing the charge of fuel and air and distributing this mixture to the cylinders through the intake manifold. The mixture must be rich enough to ensure that the cylinders located farthest away from the carburetor get enough fuel. The others, consequently, tend to run overrich. That's a big handicap for both fuel economy and emission control.

Carburetor principles: Nothing is measured, but fuel metering maintains a nearly constant proportion with airflow, due to its dependence on manifold vacuum. The sketch also makes the point that mixture for all cylinders is prepared together, upstream of the intake manifold.

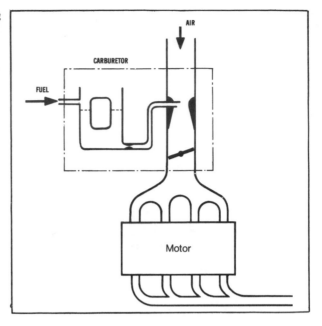

Fuel injection principles: The basic difference with a popular, modern Bosch injection system from the typical carburetor is that air and fuel are measured separately. Fuel metering takes account of mass airflow, engine speed, throttle position and engine temperature as the essential control parameters. In some systems, further parameters are added. Fuel is delivered individually to each cylinder, in uniform quantities for any given set of conditions. That saves fuel and gives more power.

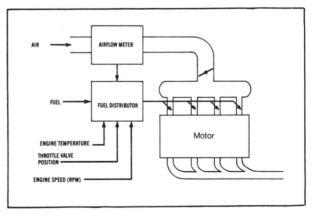

All carburetors have a throttle valve to control the volume of intake air. The mass of air that goes into a carburetor regulates engine speed and can therefore be used to regulate engine power. The throttle valve is linked to the accelerator, or gas pedal. The throttle valve is a butterfly valve that consists of a disc mounted on a spindle. The disc is roughly circular, and has the same diameter as the main air passage in the carburetor. It is located at the bottom of the carburetor, between the jet nozzle and the intake manifold. The throttle spindle is connected to the accelerator in such a manner that when the pedal is depressed, the valve opens. When the pedal is released, the valve closes. The fuel jet is fed from the reservoir and projects into the narrowest part of the venturi where air velocity is highest. Gasoline does not normally flow out of the jet nozzle but must be trickled out by the airflow.

This trickling supply of fuel is assured because there is higher pressure in the carburetor bowl than in the engine manifold. Of course, it's not practical for the carburetor bowl to be pressurized. It is left at atmospheric pressure since it is much simpler to estab-

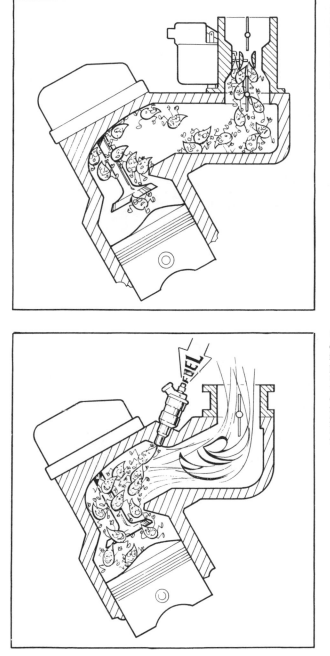

With a carburetor, fuel droplets are sucked in by manifold
vacuum for mixing with the
incoming air upstream of the
manifold flange. There is a big
risk of wetting the manifold
walls with the fuel droplets,
which upsets the balance of
the mixture.

With fuel injection, the fuel
droplets are squirted into the
inrushing airstream under
pressure. With the injector
mounted close to the intake
valve, there is little or no
chance of wetting the manifold walls. All the fuel goes
into the cylinder without delay, allowing very precise control over the air/fuel ratio.

lish a pressure difference by creating a pressure drop or partial vacuum on the engine side
of the carburetor.

But the fact remains that fuel delivery in a carburetor tends to lag behind the throttle motion, mainly due to surface tension and inertia in the fuel.

Typical fuel-injection installation. The missing carburetor means elimination of all the problems of carburetion—a major argument in favor of fuel injection. The photo shows the Bosch K-Jetronic continuous port-injection system fitted on a 2.8 liter V-6 Ford engine. The car is a British-made Granada.

Fuel shutoff: Fuel-injection systems can cut off fuel delivery completely when no power is needed, while the carburetor continues to mix fuel into the airstream in predetermined proportions. On fuel-injected cars a big energy saving can be made by a total stop on fuel delivery whenever the engine is driven by the vehicle, on a downgrade or during deceleration, with an accelerator pedal released from foot pressure. When engine speed comes close to idle speed, fuel supply is restarted to prevent stalling. At a touch of the accelerator, fuel delivery is also restored instantly.

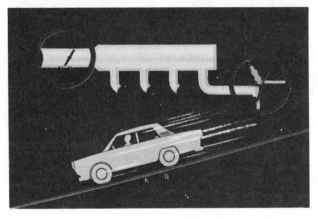

The basic carburetor operates when the throttle valve is fully open or partly open, but not when it's closed. But no driver wants the engine to stop every time he takes his foot off the accelerator. Such a car would be very tiring to drive under the easiest conditions, and almost impossible to drive in heavy traffic, with frequent stops and starts. To keep the engine running even when no power is needed, the idle circuit was added inside the carburetor. The idle jet admits fuel on the engine side of the throttle valve. Additional air is mixed with this fuel via an air bleed. The result is an entirely separate carburetor circuit which operates only when the throttle valve is closed.

Airflow through the carburetor is assured by the pumping action of the pistons. The downward movement of the piston in the intake stroke creates a partial vacuum in the cylinder. Air-fuel mixture in the intake manifold rushes in to fill the vacuum, and the gas flow set up by the pressure drop draws fresh air into the carburetor.

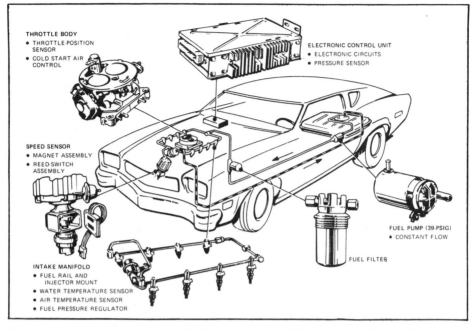

THROTTLE BODY
- THROTTLE-POSITION SENSOR
- COLD START AIR CONTROL

ELECTRONIC CONTROL UNIT
- ELECTRONIC CIRCUITS
- PRESSURE SENSOR

SPEED SENSOR
- MAGNET ASSEMBLY
- REED-SWITCH ASSEMBLY

FUEL PUMP (39-PSIG)
- CONSTANT FLOW

FUEL FILTER

INTAKE MANIFOLD
- FUEL RAIL AND INJECTOR MOUNT
- WATER TEMPERATURE SENSOR
- AIR TEMPERATURE SENSOR
- FUEL PRESSURE REGULATOR

Elements of an early-model electronic fuel-injection system. This system provides timed injection of accurately metered amounts of fuel into the port areas.

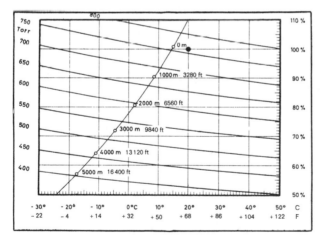

Altitude compensation: Air gets thinner as the road goes higher. The engine then gets less oxygen, and power drops off. Running with standard carburetor settings, the engine will waste fuel due to an over-rich mixture. The chart shows the relative loss in engine performance due to the effect of higher altitude. At 5,000 feet above sea level, the average carburetor engine loses about 16.5 percent of its power while burning 100 percent of the fuel. With fuel-injection systems, altitude compensation can be built into the basic controls.

A pressure drop is instigated by a device called a venturi. The venturi is a constriction in the carburetor throat. The only way a constant amount of gas can flow through a tube that gets narrower is by speeding up. And according to Bernoulli's law, an increase of air velocity through the venturi will be accompanied by a reduction in pressure. The venturi is placed so that this pressure drop will be highest in the area near the jet nozzle. Fuel will flow from the float chamber to the jet orifice because the pressure on the surface of the fuel in the float chamber is atmospheric while the jet outlet is in an area of partial vacuum. It flows out of the jet as fine spray and is carried along by the entering air.

Power output and torque curves for the same engine, fitted alternatively with carburetor and Bosch L-Jetronic fuel injection. In this case, fuel injection does not swell the low-range end of the torque curve, but gives higher torque at high speed. Power is boosted by nine percent, with the peak-power speed about 500 rpm higher.

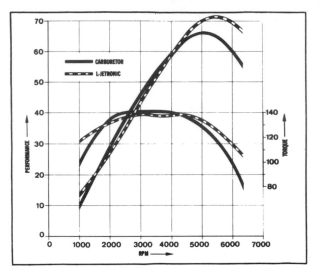

To obtain proportional and automatic fuel feed into the air stream, the carburetor has to maintain the correct air velocity past the fuel jet nozzle under light as well as heavy loads.

Venturi efficiency depends on its length-to-diameter ratio. The amount of fuel drawn into the engine depends on how big the pressure drop is. The smaller the fuel particles, as they leave the carburetor, the more easily and thoroughly will they mix with the air and vaporize in their passage through the intake manifold to the cylinder. The dimensions of the air-supply pipe and the jet are carefully proportioned to give the correct air/fuel ratio. The problem is that this ratio is not constant, and the carburetor cannot accurately adjust it to changing needs.

In any case, the mixture should be a homogeneous vapor containing no liquid fuel, for proper ignition and waste-free and complete combustion. That is conditional on thorough mixing—and this is a factor that must be taken into account also for fuel-injection systems.

With carburetors, the thoroughness of the mixing process is mainly determined by the distance from the carburetor flange to the inlet valve port, and the gas flow velocity. The rate of fuel flow at the nozzle increases faster than the pressure drop in the venturi. This means that the engine would run richer and richer as speeds rise, unless the carburetor could correct the air/fuel ratio. Air correction is achieved by introducing air into the fuel supply before it leaves the nozzle. This is usually done with an air bleed. The most common type is an emulsion tube, a short tube with holes drilled across it. It is located in a fuel well inside the carburetor. As the pressure drop increases, fuel flows faster. This lowers the level in the fuel well and uncovers more holes in the emulsion tube. As a result, additional air is bled into the mixture, and the formation of an overrich mixture is prevented.

The degree of atomization varies greatly with changes in engine speed and load (as measured from the throttle opening). A large-diameter venturi would be best for full-power operation. A small-diameter venturi would be best for part-throttle operation. It also offers fuel-economy advantages, and while it tends to provide adequate acceleration potential, it inevitably entails a slight loss in top speed.

Many manufacturers have increased venturi size to provide the full airflow needed by their engines for maximum power output. Then it was found that there was a practical limit to maximum venturi size for acceptable low-speed vehicle operation. This limitation led the industry to the modern two- and four-barrel carburetors.

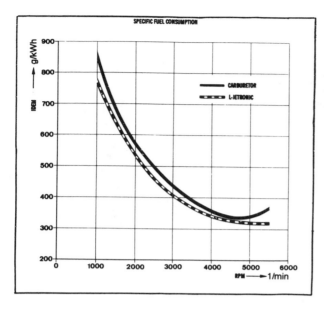

SPECIFIC FUEL CONSUMPTION

CARBURETOR
L-JETRONIC

Lower fuel consumption: In a dynamometer test of two otherwise identical engines, different in being equipped with carburetor on one and fuel-injection on the other, Bosch technicians measured specific fuel consumption throughout the speed range. The fuel-injected engine shows a clear advantage at all practical speeds.

'Barrel' is a popular term for carburetor throat. There is one venturi in each throat. A two-barrel carburetor has a primary venturi for part-load running and a secondary venturi for full-throttle running. The four-barrel carburetor has two primary and two secondary venturis. Only the primary venturis are used under part-load conditions. The throttle valves on the primary throats are linked to the valves in the secondary throats. When the primaries are, for instance, half open, the secondaries only begin to open. At full throttle, all throats are fully open. This assures a certain economy under light-load operation as well as maximum gas flow and correct air-fuel mixture for full-power operation.

Theoretically, the ideal air/fuel ratio is 14.7 parts of air to one part of fuel. However, experience has shown that it's true only in steady-state operation, such as turnpike cruising, with minute variations in throttle-valve angle and engine revolutions per minute.

For part-throttle, light-load operation, a lean ratio of 16.0 to 16.5:1 will be adequate. It must be enriched for speeding up, approaching 12:1 for full-throttle acceleration. A ratio of 10 or 11:1 will be best for a hot engine at idle. But for cold-starting, the air/fuel ratio must be as rich as 3 or 4:1.

Air and fuel don't mix very well in cold weather. Only the lightest portions of the gasoline help form a combustible mixture. For that reason, the mixture must be richer when the engine is cold and ambient temperature is low.

This is achieved by a choke mechanism. The choke is a special valve placed at the mouth of the carburetor so that it partially blocks off the entering air. By severely reducing air supply, vacuum is drastically increased at the venturi, causing the fuel flow to speed up. This results in a richer mixture. The engine starts—but will it keep running? Blocking off part of the air intake tends to stop the engine, doesn't it? Yes. When the engine is cold and operating with the choke valve shut, the total volume of entering mixture is so small that the engine has a tendency to stall. Something had to be invented to keep it running.

This much-needed invention is a simple thing called the fast-idle cam. This is a part of the choke mechanism, hooked up to the throttle linkage to give a fast idle as long as the choke is in function.

The *manual* choke is a knob on the dash, usually a push-pull. The main thing to know about it is to remember to push it back in when the engine has reached operating temperature.

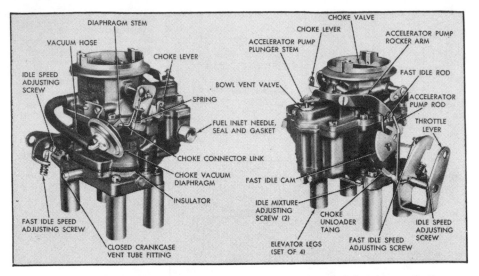

Simple and inexpensive? The simplicity and low cost of the carburetor is largely a myth. Even the simplest of venturi-type carburetors have plenty of complications.

The four-barrel carburetor is a maze of mechanical, pneumatic and hydraulic (fluid-flow) systems interacting to satisfy the engine as well as the carburetor principle will allow. Some fuel-injection systems may be cheaper.

Automatic chokes rely on engine heat for their operation. The choke valve is operated by a thermostat controlled by exhaust heat. With a cold engine, the choke valve will be closed for starting. As the engine warms up, the exhaust heat will gradually open the choke valve.

The thermostat consists of a heat-sensitive spring. The most common type is a bimetallic coil spring, much like a watch spring in appearance. This spring is fixed at its center and attached to the choke valve by a linkage at its circumferential end. The spring is contained in a housing. The housing is connected via a tube to the exhaust manifold. When the engine is cold, the spring contracts to close the choke valve. As soon as the engine begins to warm up, heat from the manifold makes the spring unwind itself, and the choke valve gradually opens.

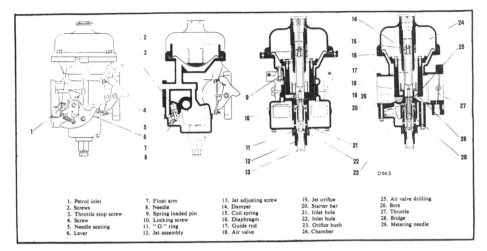

1. Petrol inlet	7. Float arm	13. Jet adjusting screw	19. Jet orifice	25. Air valve drilling
2. Screws	8. Needle	14. Damper	20. Starter bar	26. Bore
3. Throttle stop screw	9. Spring loaded pin	15. Coil spring	21. Inlet hole	27. Throttle
4. Screw	10. Locking screw	16. Diaphragm	22. Inlet hole	28. Bridge
5. Needle seating	11. "O" ring	17. Guide rod	23. Orifice bush	29. Metering needle
6. Lever	12. Jet assembly	18. Air valve	24. Chamber	

The air-valve or constant-vacuum carburetor: The whole air-valve assembly rises or falls according to the strength of the vacuum in the depression chamber (24). The piston (18) controls the cross-sectional air inlet area, its motions being damped to avoid overreactions to rapid flooring of the accelerator pedal. One air-valve carburetor has only one air passage, and for proper mixture distribution, a multicarburetor setup may be required for multicylinder engines.

The intake manifold incorporates a special passage for exhaust gas to warm the incoming air-fuel mixture and improve atomization on cold starts. After the engine is warm, this heating is undesirable as it would cause the fresh mixture to expand before entering the cylinder. This would reduce the volumetric efficiency of the engine. Therefore, a heat-riser valve is provided to direct exhaust-gas flow according to engine temperature.

This valve is located on the exhaust manifold. It regulates the amount of exhaust gas allowed to pass through the intake manifold. A bimetal spring attached to the control valve shaft gradually restricts the amount of exhaust-gas flow to the intake manifold by slowly closing the valve. When proper operating temperature is reached, the exhaust gas is routed to the muffler and out the tail pipe.

Prior to the advent of emission controls, fuel delivery was considered accurate enough if variations in air/fuel ratio across the full range of airflow mass and velocity were kept within plus-or-minus five percent of the nominal setting. By 1969, nothing above a three-percent variation could be tolerated, and by 1972, the margin was reduced to between 1.0 and 1.5 percent. Such precision is extremely difficult to obtain in mass-produced carburetors.

The complications of making the carburetor fit the requirements of the engine do not end there. All cars with automatic transmission have a dashpot to keep the engine from stalling when the accelerator is suddenly released. The dashpot is actuated by an arm of the throttle lever when the throttle is closed. It cushions the closing of the throttle, letting fuel metering abate gradually instead of being shut off all at once.

Automatic altitude compensation would be quite expensive to include in a carburetor. Because of rarefied air at high altitude, normal setting would give an overrich mixture. Statistically, few cars move back and forth between the two different environments of low and high altitude, and the industry has been content to deliver cars with different carburetor settings according to destination.

The carburetor, despite all its complications—air bleeds, correction jets, acceleration pumps, emulsion tubes and choke mechanisms—remains a compromise. The additional costs of better carburetor design compromises are helping to push the industry toward fuel injection.

But as we shall see later, fuel-injection systems are not, themselves, free of compromise. Two arguments remain, standing unchallenged in favor of fuel injection. One is the engine's ability to run with higher compression ratios when the fuel is injected.

The other argument—and some experts say it is the only real advantage of fuel injection—lies in the freedom of intake manifold design. With carburetors, manifolds must be designed to encourage fuel atomization, and discourage raw fuel droplet accumulations on the runner surfaces. To this end, manifold heating must be provided.

For engines with carburetors, the intake manifold has two main duties:
1. It must deliver equal quantities of fuel and air mixture to all cylinders.
2. The mixture must possess the same chemical and physical characteristics in all cylinders.

Gases are both elastic and viscous. With high gas velocities through imperfect passages the gas composition will break up and gas flow will change from smooth to turbulent. Only individual particles will maintain theoretical velocity.

When the cross-sectional area of the manifold is too large, gas velocity drops below a critical point where drops of raw fuel begin to settle on the manifold walls. This changes the proportion of air to fuel as metered by the carburetor.

The manifold cross-section is usually as large as possible for ideal power conditions, without reaching the critical point. Inner walls of the manifold are made as smooth as possible because gas flow separation can be caused by roughness in the surface. On the other hand, this tends to promote fuel droplet deposition. Sharp corners in a manifold are sometimes deliberately built in. They can cause a film of fuel to tear loose from the wall and re-enter the air stream, but since this occurs at random, it can also cause a momentary upset of the air/fuel ratio.

Depending on the exact location of the injector nozzles, the mixture-control considerations do not apply to manifolds for fuel-injected engines. With fuel injection, the engineer has a free hand to exploit ram effects to the maximum.

Ram is an air momentum effect. Once gas flow has been started and accelerated, there is inertia in the gas when its flow is suddenly stopped by a closed valve. This can be utilized to set up pulses in the manifold runners that will force more gas past the intake valve than would be possible by suction only. Intake pipe length governs ram effect. There is an ideal length for each cylinder for each engine speed, and all manifolds are a compromise between optimum high-speed and best low-range gas flow characteristics.

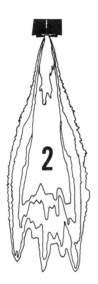

Fuel Properties

When we say 'fuel injection,' in comparison with carburetors, the fuel is, of course, the same: gasoline. Fuel injection is used on all diesel engines, but diesel engines operate by different principles, and their injection equipment and fuel properties will not be discussed here.

Fuel injection is fully compatible with gasohol and certain other alternative or future fuels; but for the moment, existing systems are designed and made for gasoline only. The following discussion will be restricted to gasoline as you buy it at the pump.

The first fact that needs to be faced about this fuel is that raw gasoline will not burn. It must first be changed from liquid to vapor and supplied with an adequate volume of air to assure that enough oxygen to support combustion enters the engine. Gasoline is a hydrocarbon, made up of about fifteen percent hydrogen and eighty-five percent carbon. There are other ingredients that we'll take a closer look at later, but this is all you have to know about gasoline to understand its use as motor fuel. Air is a mixture of twenty-one percent oxygen, seventy-eight percent nitrogen and one percent other gases. But only the oxygen combines with the gasoline.

Before gasoline can be thoroughly mixed with air in proper proportions, it must be broken up, or atomized, into finely divided particles. But the engine won't run satisfactorily if it only gets an air-fuel mixture of the same proportions all the time. The mixture must be adjusted to speed, load conditions and temperature. And the total volume as well as the air/fuel ratio must be continuously adjusted to these conditions.

Gasoline is obtained by refining crude oil. Crude oil as it comes from the ground is a mixture of thousands of different chemicals which range from extremely light gases to semi-solid carbon-containing materials such as asphalt or paraffin wax. The gases are dissolved in the other components of the crude oil because of the extreme pressure at which petroleum is stored in the ground.

Water weighs 8.33 pounds per gallon. Crude oils vary between 6.5 and 8.3 pounds per gallon. As a liquid, it may be as thick and black as melted tar or as thin and colorless as water. Its characteristics depend on the particular oil field from which it comes. Crude oils contain sulfur compounds in varying amounts. Sulfur is undesirable in motor fuel because it gives rise to bad odors, is corrosive to engines, poisonous to some catalysts and may reduce the efficiency of antiknock compounds. A maximum amount of sulfur is removed as part of refining.

The refining process begins with distillation of the crude oil. That is followed by vacuum fractioning which separates light and heavy oils, gas oil and bitumen. Refining also includes such supplementary operations as the stabilization of the gasoline to eliminate condensible elements, refining of white spirits, extraction of oils by means of solvents and the processing of paraffin and bitumen.

The cracking process is also important. Higher-boiling hydrocarbon molecules can be broken down or 'cracked' into lower-boiling ones by subjecting them to extreme temperatures. Thermal cracking was the natural and simple way to do this. However, the original method has been almost completely replaced by catalytic cracking which gives greater freedom of production for a refinery.

Cracking is fractional distillation combined with other operations. After an initial stage of physical purification, the oil is subjected to fractional distillation at atmospheric pressure in columns some thirty meters high. They are heated at the base, and temperatures range between 350°C and 70°C. The most volatile products run to the top. From top to bottom we have: gasoline, kerosene, gas oil, fuels and lubricants.

Catalytic cracking is a reforming process which transforms heavy essences of first distillation into light ones with a higher octane number. In catalytic reforming, there are molecular rearrangements as well as molecular splitting. The performance of gasoline is determined mainly by its volatility (tendency to boil, and vapor pressure), by its anti-knock quality and by its cleanliness and stability. These characteristics are determined by the refiner's selection and processing, plus the additives and inhibitors the refiner adds.

Gasoline must form vapors at a low temperature to assure easy starting. It must vaporize at an increasing rate as carburetor and manifold temperatures rise to assure fast warmup, smooth acceleration and even fuel distribution among the cylinders. The vaporizing characteristics must be in keeping with the climate and altitude where the gasoline is to be used, to prevent vapor lock and fuel boiling inside carburetors, pumps and fuel lines.

Gasoline should contain few extremely high-boiling hydrocarbons to insure good fuel distribution and freedom from crankcase deposits and dilutions. A high antiknock quality (octane number) throughout its boiling range is needed to assure freedom from knock at all engine speeds and loads. Gum content must be low to prevent valve sticking, carburetor deposit difficulties plus deposits inside engine and intake manifolds. Gasoline must also have good stability against oxidation to prevent deterioration and gum formation in storage.

The antiknock property of a gasoline is indicated by its octane number. The octane scale was created by giving the number zero to heptane (C_7H_{16}) and the number one hundred to iso-octane (C_8H_{18}). Numbers between zero and one hundred indicate the proportion of each if the two are mixed. The octane number of a gasoline is determined by a test comparing it with a mixture of heptane and iso-octane. If the gasoline shows the same tendency to knock as a mixture containing six percent heptane and ninety-four percent iso-octane, its octane number is 94.

There are two ways of establishing a fuel's octane number. *Research method:* The test engine is run under closely controlled conditions of speed, air intake temperature and ignition timing. *Motor method:* The test engine is run with variations in speed, air intake temperature and ignition timing. The difference between Motor and Research method ratings for the same fuel is called 'sensitivity.'

It is impossible to obtain more than one-hundred percent iso-octane in a reference fuel blend, but some fuels have ratings above one-hundred octane. Then the reference fuel becomes iso-octane plus a certain amount of tetraethyl lead. The knock value of such fuels can be defined as milliliters of tetraethyl lead per gallon of iso-octane.

The fuel-injection specialist must take close account of an engine's octane requirement. Octane requirement can be defined as the octane number that produces a given level of knock resistance during acceleration with a throttle opening that is known to produce maximum knock.

Octane requirement depends on many variables. Identical vehicles coming off the assembly line can have differences of up to ten octane numbers. The conditions that dictate octane requirement are: atmospheric pressure, relative humidity, air temperature, fuel characteristics, air/fuel ratio and variations in air/fuel ratio between individual cylinders in the same engine, oil characteristics, spark timing, distributor advance curve, variations in timing from one cylinder to another, intake manifold temperature, water jacket temperature, condition of coolant antifreeze, type of transmission and the presence of hot spots in the combustion chambers.

Antiknock compounds are usually lead-based. Since lead is a poison, the Environmental Protection Agency has imposed strict limits on the lead content of gasoline. Some European countries have no lead restrictions. Sweden and West Germany have limits for lead content, but less severe than those of the U.S.A. and Japan.

Where leaded fuels are still produced, scavengers are added to remove the small amount of lead compounds that might remain in the engine after combustion. These scavengers include bromine and/or chlorine. They convert the lead compounds to lead bromine and lead chloride. These are gaseous at the temperatures prevailing inside the engine. The scavengers remove practically all lead and dispose of it in the exhaust fumes. When lead is found in the combustion chamber deposits, it is mainly because organic residues from burned oil or fuel have acted as blinders for them.

Antioxidant chemicals are added to gasoline to protect against formation of gum and peroxides. Numerous chemicals have been used as antioxidants. Until recently, phenylene diamine, aminophenols and dibutyl-cresol were those most widely used. Today the most used and most effective antioxidant is a new family of ortho-alkylated phenols.

Metal deactivators, usually an amine, are often used along with antioxidants. They prevent trace amounts of copper (picked up from the piping or the engine fuel system) from acting as a catalyst for the formation of undesirable materials in the gasoline. Other additives include anti-icers (alcohol) to prevent carburetor icing and fuel line freeze-up and antirust agents. There are detergents to keep carburetor parts clean—and phosphorous additives to combat surface ignition and spark plug fouling. Dyes are added to identify leaded gasoline.

Gasoline is carried in the fuel tank, usually positioned in the tail end of the car, under or ahead of the trunk floor. Fuel level in the tank is sensed by a float, coupled to a sender unit, connected to a gauge on the instrument panel.

Filtration of the fuel is necessary to keep water and dirt from entering the engine. Water is the main problem. Moisture in the air inside the tank will condense and sink to the bottom of the tank. Some cars have a filter in the tank. It is placed at the main fuel line where it leaves the tank. Some cars even have coarse filters in the filler neck.

The main fuel filter is usually integral with the fuel pump. Fuel passes through a filter screen before passing through the outlet. Dirt and water caught by the screen fall to the bottom where they can be removed.

Another type of filter is made of a series of laminated discs placed within a large sediment bowl. This bowl acts as a settling chamber for the fuel and encloses the discs or strainer. Fuel enters the filter at the top and flows downward. It flows between the discs and then up a central passage to the outlet connection at the top. Dirt and water cannot pass between the discs because the clearance is too fine. Often, the carburetor carries a separate filter to clear the fuel before admission to the float chamber.

There are two common types of fuel pumps: mechanical and electrical. The mechanical pump has been common on U.S.-built cars since 1926, but it is gradually being replaced by electric pumps. Electric pumps have long been popular in Europe.

The pump action is provided by a spring-loaded diaphragm inside a housing. An inlet valve admits fuel to the housing but blocks its return. The diaphragm exerts pressure on the fuel inside and forces it up the outlet valve, which also assures that it doesn't return into the pump. The drive is taken from an eccentric on the camshaft, which sets up a rocking motion in the pumping arm.

The action of the arm pulls down the hub of the pump diaphragm. This creates a partial vacuum which allows atmospheric pressure, acting on the surface of the fuel in the tank, to push fuel along the line and fill the pump housing. The pump arm or lever is divided in two parts at its pivot. The joint is so arranged that the part operated by the engine carries the other part around only when it is moving in a counterclockwise direction. The other part is attached to the diaphragm.

The right-hand portion of the lever is spring-loaded against the eccentric. It will continue to turn clockwise as the eccentric continues to revolve and leave behind the left-hand portion of the lever. Subsequent movement of the diaphragm is determined by whether or not the carburetor is full of fuel. If it's full, the needle valve will prevent more fuel from entering the float chamber. The diaphragm will be kept at the bottom dead center of its travel. The right-hand portion of the lever will continue to oscillate without effect.

If the float chamber is not full, the diaphragm will be pushed upward by its spring and force fuel into the carburetor. The left-hand portion of the lever will catch up with the right. On the next rotation of the eccentric, it will again be pushed down in readiness for the next charging phase.

The electric pump is usually mounted away from the engine. Common locations are on the cowl or near the tank. A few inside-tank applications exist. The diaphragm is operated electromagnetically. An iron armature attached to the diaphragm is drawn toward an electromagnet when the latter is energized. A spring tends to hold the diaphragm in the 'pump empty' position. Having reached this position, an extension rod attached to the armature closes two contacts and allows an electric current to flow through the magnet coil. The armature is therefore pulled toward the magnet, and the retraction of the diaphragm brings fuel into the pump.

When the armature reaches the left end of its travel the extension rod operates a throw-over mechanism which separates the contacts so that the armature and diaphragm can return to the right. The diaphragm and armature, under spring pressure, move right depending on the quantity of fuel the carburetor is ready to accept. No hand-priming device is required. The pump begins to operate as soon as the ignition is switched on.

The main advantage of the electrical pump is its immunity to vapor lock. That's a phenomenon caused by vaporization in the fuel line or pump, which means a gas is being pumped instead of fuel, with the result that the engine stalls. With electric pumps, fuel is pushed through regardless of under-hood temperature.

Since 1970 in California and 1971 nationwide, cars with carbureted engines have had to be equipped with evaporation control systems. It was found that evaporation from the carburetor float bowls and fuel tanks could cause up to twenty percent of the total hydrocarbon output of a car.

These systems worked by trapping the vapors in a charcoal canister installed under the hood, and emptying it into the carburetor when the engine was restarted. Fuel tanks were provided with sealed caps to prevent vapor leaks. The regulations concerning evaporation control gave the industry added impetus to investigate fuel injection.

The heat value of gasoline can be directly converted into useful energy of work. Each BTU (British Thermal Unit) of heat energy can be considered equivalent to 778 foot-pounds of work. One gallon of gasoline is capable of developing about eighty-nine million foot-pounds of work, which is equivalent to 2,700 horsepower for one minute, or forty-five horsepower for one hour. The higher the amount of heat energy that is converted into useful motion, the higher the engine's thermal efficiency.

Much of the fuel is wasted because of the complex process of making a car go. First, chemical energy is converted into heat energy and gas pressures. The pistons and cranks change these elements into mechanical rotation, and additional machinery finally uses this rotation to produce dynamic motion.

Present-day gasoline engines reach a thermal efficiency of about twenty-seven percent. This means that twenty-seven percent of the heat value of the fuel is converted into useful energy. The cooling water carries off about thirty percent of the heat value, and the exhaust gases blow away about thirty-three percent. Driving the cooling fan and water pump costs about three percent.

Friction losses are generally a function of engine speed. Most friction losses also vary with changes in load. An increase in piston thrust increases cylinder friction while higher-bearing loads, largely owing to distortion, increase shaft friction. On an average, friction losses account for three percent of the heat energy.

On an average, pumping losses account for four percent of the heat energy. Contrary to what you might expect, pumping losses are not proportional with engine speed. The exhaust stroke, for instance, is negative work, since there is a load on the piston crown. The intake stroke represents the bulk of the pumping losses. These losses are at maximum with smallest throttle opening, and diminish when the throttle is opened.

Combustion and Thermal Efficiency

The process of mixture preparation, accomplished variously by carburetors and fuel-injection systems, has a vital effect on combustion and thermal efficiency. Just think of the injected engine's ability to run with higher compression ratios, and the importance of precisely controlling the air/fuel ratio becomes crystal clear. It is also an essential part of controlling exhaust emissions.

In terms of fuel savings, there is enormous promise in engines that will accept leaner mixtures. Now that every fraction of a percentage point in fuel savings is the subject of intensive laboratory investigations, the possibility of igniting and burning very lean mixtures keeps thousands of research engineers and technicians at work in the automobile industry, its supplier industries and the petroleum companies.

This avenue of research is not totally concentrated on fuel-injected engines, but the broad thrust is aimed at some system of fuel injection. Why? Because of the assumption that greater accuracy can be obtained under any and all combinations of operating conditions.

If you accept the basic truth that the energy-efficiency of an engine increases as the air/fuel ratio becomes leaner and leaner, you end up with the conclusion that maximum efficiency is not reached until the engine is running on one-hundred percent air and not a drop of gasoline. Since one-hundred percent of the energy comes from the fuel, the conclusion cannot be correct. In practice, the engine won't run on air alone.

Just how lean an engine will run reliably and with 'clean' exhaust varies continuously according to operating conditions. Consequently, the air/fuel ratio must be under continuous control with fast response to a maximum of information about temperature, speed, load and so on.

It is desirable, for fuel-economy reasons, to run the engine as close to the lean limit as possible. But crossing this limit has dire consequences, far more dangerous than erring on the rich side.

There is risk of early flameout with overlean mixtures, which would cause an increase in hydrocarbon (HC) and carbon monoxide (CO) emissions. Leaning-out the mixture even more will lead to the situation where ignition cannot occur, and the engine will stall.

For successful operation close to the limits of reliable running and meeting emission standards, the very highest precision is needed for adjusting the mixture preparation. The most attractive approach is a fuel-injection system with electronic control.

Combustion starts toward the end of the compression stroke and lasts till about two thirds of the expansion stroke has been completed. The consistency of the charge and the timing of the spark are as important as the shape of the combustion chamber.

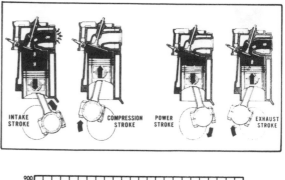

INTAKE STROKE COMPRESSION STROKE POWER STROKE EXHAUST STROKE

Theoretical pressure/volume for a four-stroke cycle engine: The tip at lower right represents the intake stroke. Following the lower curve leftward, we see that compression brings a temperature rise. The pressure is increased as cylinder volume is diminished, and skyrockets when the mixture is ignited. Peak temperature is reached before the point of maximum pressure. Then the pressure and temperature decline during the power stroke until the exhaust valve opens. Engine design compromises resulting in heat and friction losses keep the curve theoretical—but fuel-injection engines come closer than carbureted engines.

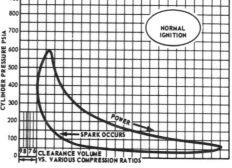

To obtain complete combustion of all the hydrogen and carbon of one pound of gasoline, we need the entire oxygen content of 14.7 pounds of air. This air/fuel ratio of 14.7:1 is called stoichiometric, which more or less means 'ideal.'

Under the right conditions, this mixture, when ignited, will release the total heat energy stored in the fuel without leaving any residual oxygen or unburned hydrocarbons. The combustion products are nontoxic, nonpolluting and harmless: carbon dioxide (CO_2) and water vapor (H_2O).

Mixtures with more than 14.7 parts of air to one part of gasoline are called *lean*. If there is less air, the mixture is called *rich*. Engineers have another ratio when they discuss rich/lean mixtures. It's the equivalence ratio. An equivalence ratio of one corresponds to the stoichiometric ratio.

The equivalence ratio is symbolized by the Greek letter Lambda. In everyday terms, Lambda can be expressed as the ratio between the actual amount of air delivered and the theoretical air requirement (stoichiometric).

With Lambda equal to 1.0, the two are identical, and the engine is receiving a stoichiometric air-fuel mixture. With Lambda values less than 1.0, we are dealing with an air shortage and, therefore, a rich mixture. With Lambda values higher than 1.0, we are dealing with an air surplus and, therefore, a lean mixture.

Each engine has its own lean limit, and that's why there is no fixed general number for the truly best economy figure. But it has been determined that maximum efficiency for any real engine is reached with at least half the fuel of the stoichiometric ratio, or below a Lambda value of 2.0.

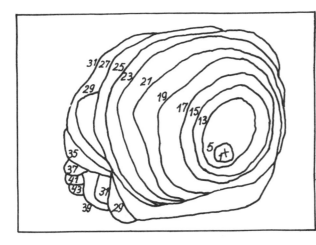

Flame front travel: The flame front spreads from the spark plug (x) and by filming the combustion process, through a 'window' and using a mirror, it has been possible to map the progress of the flame front. This is from a Mercedes-Benz engine with parallel, canted overhead valves and fuel injection. The numbers indicate the sequence in milliseconds, revealing a critical slowing-down in the quench area adjacent to the inlet valve.

The car engine has another problem in that, even when supplied with fuel mixture in a stoichiometric ratio, its operating parameters may not allow time for complete combustion. Consequently, even a charge that is theoretically in perfect balance, will normally leave some unburned fuel in the form of raw hydrocarbons and carbon monoxide. These are expelled in the exhaust (which is treated by a catalyst [on most post-1975 cars] that breaks them down chemically into harmless constituents before letting them escape into the atmosphere).

Near the lean limit, hydrocarbon and carbon monoxide emissions are at their minimum. Nitrous oxide emissions are highest at stoichiometric, and fall off toward both the rich and lean ends of the curve.

The fuel-injection specialist must have an acute awareness of the most intimate goings-on inside the engine if the injection system is to come anywhere close to meeting the lean-burn requirements of the engine. Thanks to the existence of advanced materials and instrumentation, research engineers can now film the combustion process in laboratory engines with ultra-high-speed cameras, and obtain the vital data for planning, laying out, installing and setting the fuel-injection system.

Normal combustion occurs when the mixture in the combustion chamber is ignited by a spark plug firing at a pre-set point, starting a wave of flame spreading out from the spark plug. It's like burning grass in a field. Just as it takes time for the flame to move across the field, it takes time for the flame front to travel across the combustion chamber. This flame front continues to move across the combustion chamber until it reaches the other side. The compressed mixture burns smoothly and evenly.

Flame-front speed varies from twenty feet per second to over 150 feet per second. This speed depends mainly on air/fuel ratio, compression ratio, turbulence and combustion chamber design. Flame travel is quite slow when the mixture is too rich. It is also slow when the mixture is too lean. The faster the flame front travels, the smaller the risk of abnormal combustion.

High compression gets more energy out of the fuel charge. The pressure gain in a cylinder with an 8.5:1 compression ratio is about 500 psi (pounds per square inch). The pressure gain in a cylinder with a 10:1 compression ratio is about 750 psi. It is true that more work is spent in compressing the mixture when the compression ratio is high, but there is a large gain in thermal efficiency and power because less heat is rejected to the cooling system and more of the caloric energy in the fuel is put to useful work.

Normal pressure rise is 3.5 to 4.0 times the initial pressure. Pressure-rise rates can reach 20 psi per degree of crankshaft rotation. If the initial pressure at the moment of firing is 160 psi, the peak pressure during combustion will be almost 600 psi.

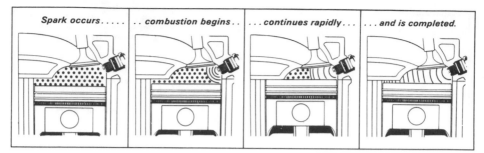

Normal combustion: The procedure shown in this sequence applies equally to carbureted and fuel-injected engines.

These sketches show preignition in a typical carbureted engine. It can also occur with fuel injection, being mainly dependent on combustion-chamber design.

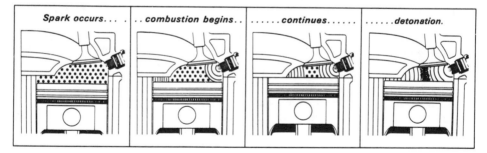

Detonation can occur with carbureted and fuel-injected engines alike, but with fuel injection, the risk is smaller. Consequently, fuel-injected engines are able to operate with higher compression ratios, which improves thermal efficiency and saves energy.

Cylinder pressures are highest at wide-open throttle, and lowest during low-speed cruising or idling conditions. Under high pressure, combustion is speeded up. Turbulence in the combustion chamber will also speed up the flame front. Turbulence means a swirling airflow pattern. Turbulence is designed into an engine. All types of combustion chambers can be given some amount of turbulence. Too much turbulence, on the other hand, is undesirable because it will increase the heat loss to the cylinder walls.

Conditions in the combustion chamber eventually determine the degree of success or failure of the fuel-injection system. Without a thorough map of those conditions, the fuel-injection specialist can do nothing for the car manufacturer.

What we demand from a fuel-injection system can be summed up in six main requirements:

1. It must give accurate control of the air/fuel ratio under all conditions. That means that it must be able to produce the mixture that will enable the engine to reach full power at full throttle, and to deliver a mixture that will let the engine operate with maximum thermal efficiency under part load.

2. It must assure accurate control of the fuel distribution so as to provide a uniform mixture in all cylinders.

3. It must not permit wetting of the intake manifold walls with raw fuel, for that would upset the mixture control.

4. It must assure adequate atomization of the fuel, so that a correct mixture is present in each cylinder at the moment of ignition.

5. It must not complicate the maintenance schedule nor introduce elements that could compromise the overall reliability or life expectancy of the car.

6. It must be mass-producible at moderate cost, and preferably have a cost advantage over a complete carburetion-cum-emission-control system.

Fuel-Injection History

Fuel injection has come a long way in the past twenty-five years. As an invention, however, its history goes back to the early days of the carburetor. And just as the best reasons for using fuel injection today are to be found in the shortcomings of the modern carburetor, it was the lack of refinement and versatility in ancient carburetors that prepared the way for the first fuel-injection experiments. The origins of fuel injection are inseparable from early carburetor history and the evolution of motor fuels.

Carburetor science began in 1795 when Robert Street achieved evaporation of turpentine and coal-tar oil in an atmospherical engine—one working without compression of the charge.

But it was not until 1824 that Samuel Morey and Erskine Hazard created the first carburetor (also for an atmospherical engine). Its method of functioning included preheating to promote vaporization. Morey was an American inventor, and Hazard an English patent attorney.

About 1825 Michael Faraday was making experiments with the vaporization of liquid hydrocarbon fuels. By distillation of mineral oil, he discovered benzole, which he called bicarbonate of hydrogen.

In 1833 the German chemical professor Eilhard Mitscherlich at the University of Berlin was able to obtain thermal splitting of benzo-acids. The new product was a fuel which he called Benzin (gasoline in American terminology, petrol to the British).

In 1838 William Barnett, an English mechanic, was granted a patent for a device to vaporize gasoline. This invention was intended for use on the compression-type engine Barnett was experimenting with.

A pistonless, atmospheric engine built in 1841 by the Italian scientist Luigi de Cristoforis, was equipped with a surface carburetor in which an airstream directed over the fuel tank was made to pick up fuel vapors.

The American doctor, Alfred Drake, conducted experiments with combustion engines, trying to use gasoline instead of gas. In the process, in the years 1848-50, he made several types of carburetors.

A French inventor named Mille received a patent in 1859 for a gas-producing device that could qualify for the description of carburetor.

The inventor of the Deutz four-stroke gas engine, Nikolaus August Otto, began experiments in 1860 with a combustion engine having a device for vaporizing liquid hydrocarbon fuels. He tried it out with white spirit but had no success with that type of fuel.

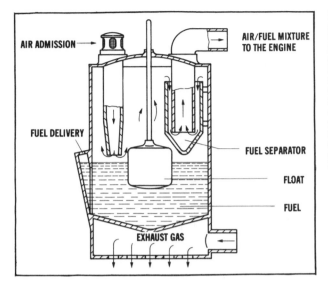

AIR ADMISSION →

AIR/FUEL MIXTURE
TO THE ENGINE

FUEL DELIVERY

FUEL SEPARATOR

FLOAT

FUEL

EXHAUST GAS

Principle of the surface carburetor: A float chamber heated by exhaust gas provided a favorable environment for evaporation. Above the fuel level, air entered via a duct controlled by a rotary valve, circulated above the surface long enough to get saturated with gasoline fumes and then the mixture was sucked into the engine via a duct carrying a separator which returned excess fuel to the float chamber.

In 1865 Siegfried Marcus of Vienna, Austria, applied for a carburetor patent, whose text stressed the simplicity of his device, compared with the costly and complicated vapor generators then in existence.

In 1867 both Étienne Lenoir and Otto displayed gas engines at the World's Fair in Paris. However, Lenoir also showed a petroleum-fuel engine with a carburetor that seems to have been overlooked at the time. Marcus was back in 1870 with a petroleum engine, working on the same principles as Otto's (Gasmotorenfabrik Deutz).

The discovery of a new basic principle occurred in 1873 when Julius Hock, working in Vienna, built an atmospheric petroleum engine equipped with a primitive form of spray carburetor.

When George Bailey Brayton began production of engines in Boston in 1874, they were equipped with surface carburetors of his own design.

It was in 1875 that Wilhelm Maybach of Gasmotorenfabrik Deutz first converted a gas engine to run on gasoline. He was in the shop one day when he suddenly had the idea of shutting off the gas, and seeing what would happen if he just held a rag, wetted with gasoline, at the manifold entry. The engine ran on until the rag was nearly dry.

A historically significant sketch of an improved surface carburetor was actually drawn by Otto's partner, Eugen Langen, for a business associate from Brussels. It is not known if the invention stemmed from Otto or Maybach, but in any case, Deutz engines from 1876 onward were equipped with this new device.

In 1880 Siegfried Marcus introduced a brush-type carburetor for his four-stroke petroleum engines, and two years later he applied for a patent from the patent office in Berlin.

The invention consisted of a circular brush mounted on a shaft, and during its rotation, the bristle came in contact with a metal plate. The resulting interaction shook fuel droplets loose from the brush and projected them into the air, where they evaporated.

In 1883-84 Marcus built a new four-stroke petroleum engine, using a brush-type carburetor with metered delivery of a fuel mist.

A wick carburetor was first used for the motor carriage (whose engine disintegrated in its first test) built in 1883 by Edouard Delamare-Deboutteville and Leon Malandin in Fontaine-le-Bourg, France.

Principle of the wick carburetor: The double-ended wick,
suspended on a rod, was im-
mersed in gasoline in a chamber linked to the float chamber. Due to wick porosity, fuel
permeated it, and the incoming air picked up fuel from its
upper portion.

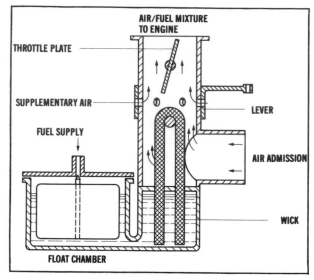

The wick carburetor, as later made popular by Lanchester and others, was a static type of device in which a wick soaked up the fuel on its submerged part and yielded it to the air on its exposed part.

A watchmaker in Munich, Christian Reithmann, obtained a patent for an 'appa-ratus to produce working gas for engines.' It was a wick carburetor, built in 1884.

In Turin, Italy, at the same time, Professor Enrico Bernardi displayed an industrial sewing machine driven by a gasoline engine having a carburetor.

Fernand Forest, a prolific French mechanic and inventor, who had built a rotary-valve four-stroke engine as early as 1871, built a new engine in 1884, using a carburetor with a float chamber and fuel jet.

By 1885 Otto obtained the results he had aimed for but failed to get in 1860, with a variety of liquid hydrocarbon fuels, including gasoline and white spirit, using an im-proved surface carburetor.

Karl Friedrich Benz used a surface carburetor of his own, designed for his first car of 1885. In the Benz system, air was drawn through the fuel tank in a controlled pattern with several reverse bends. The fuel tank was heated, so that the air, as it expanded, be-came saturated with vaporized fuel.

Benz's surface carburetor had the drawback that mixture control was haphazard at best, and under most conditions, it failed to vaporize sufficient amounts of fuel.

By that time, however, Wilhelm Maybach had completed his tests with the float-chamber carburetor, which was ready for patenting and production. Like Benz, Maybach was also making use of exhaust heat to preheat the air charge.

In the fall of 1886 Karl Benz improved his carburetor with the addition of a float valve to assure constant fuel level.

Wilhelm Maybach laid out the spray-jet carburetor in 1892, which became the basis for all subsequent carburetors. Maybach never stopped looking for better ways to mix air and fuel, and in 1894 applied for a patent on his improved spray carburetor. In it fuel was delivered in a showerhead pattern from a jet nozzle supplied from a constant-level float bowl.

In 1901 an American named Krastin perfected a two-barrel carburetor, for which he claimed consistently 'good' mixtures regardless of mass airflow.

Arthur Krebs, technical director of Panhard et Levassor in Paris, invented a three-part carburetor in 1902, with automatic air bypass to minimize deviations from the ideal

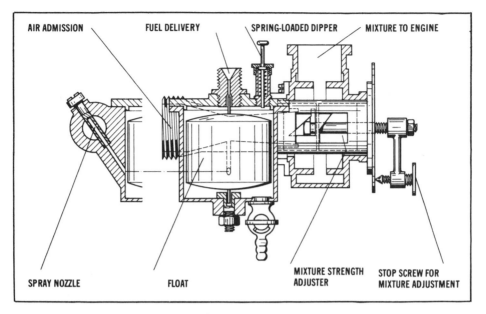

AIR ADMISSION   FUEL DELIVERY   SPRING-LOADED DIPPER   MIXTURE TO ENGINE

SPRAY NOZZLE   FLOAT   MIXTURE STRENGTH ADJUSTER   STOP SCREW FOR MIXTURE ADJUSTMENT

Maybach's spray nozzle carburetor introduced a fuel-feed pipe from the float chamber directly to the mixing tube, which was provided with manual adjustment for mixture strength (lean/rich).

Maybach spray nozzle carburetor became refined as the Phoenix carburetor and inspired Arthur Krebs of Panhard et Levassor to introduce further improvements. It set the pattern for all fixed-venturi carburetors for generations.

air/fuel ratio with increasing gas-flow velocities. Krebs used manifold vacuum to open a valve and admit supplementary air.

It was in 1905 that George Skinner patented the constant-vacuum (air valve) carburetor. The S.U. (Skinner Union) became popular, and the principle is in use today by Zenith-Stromberg as well as S.U. (now part of British Leyland).

Between 1898 and 1901, Gasmotorenfabrik Deutz built three hundred stationary engines with low-pressure fuel injection into the intake ports. The fuel was kerosene, and the injection equipment included a plunger pump with pressure-control valves.

The prior art included a patent for a compressed-air metering device issued to a Frenchman named Eteve in 1881, and a German patent issued to J. Speil in 1883 for a method of injecting fresh fuel into a flame-filled chamber linked to the cylinders.

Opening the throttle plate (1) in an SU carburetor causes the airflow through the venturi to speed up. Higher velocity lifts the piston (2) and tapered needle (3), which admits more fuel through the main jet (4) from the float bowl line (9). Jet size and throat diameter set limits on carburetor capacity.

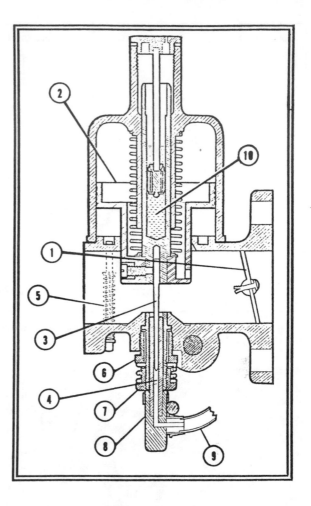

In Paris, the fabled inventor Henri Tenting is reported to have conducted experiments with fuel injection as early as 1891. He was not motivated by knowledge of the shortcomings of the modern carburetor, but seems to have made an attempt at simplification of the process of mixing air and gasoline.

Tenting failed, not because of anything wrong in his ideas, but because he lacked an injection pump to time and meter the fuel supply and seals that would enable him to control fuel pressure without leaks. His ideas were far in advance of the tools and machines of his epoch.

Also about 1883, Edward Butler of Erith in Kent, England, made an engine with an injection system which forced fuel under pressure through a hollow-stem inlet valve. But he never developed this invention to the practical stage.

Man's mastery of controlled flight turned the tables for fuel injection. The carburetors of the time were unable to cope with flying maneuvers, being tricked into either starvation or flooding by severe banking, climbing and diving.

The four-cylinder engine built by Wilbur and Orville Wright for their 1903 airplane was equipped with an interesting type of fuel injection. It used a gear-type pump which delivered fuel under pressure into the intake ports.

The Antoinette engine that powered the Santos-Dumont airplane which made its first historical flight in 1906, also featured fuel injection.

Fuel-injected Deutz single-cylinder gasoline engine Type E4 with injection pump and port-mounted nozzles was built from 1897 to 1905. It delivered 30 hp at 250 rpm. Deutz later became a leader in diesel-engine design and construction.

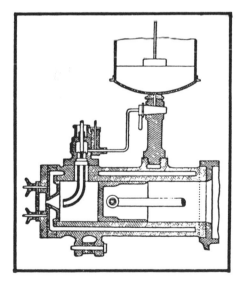

Speil's patent from 1883 shows gravity-feed from fuel tank to port-mounted injector nozzle. Fuel metering was assured by a hand-operated linkage to the valve below the tank.

The brilliant Antoinette engineer, Leon Levavasseur, had introduced calibrated injectors in addition to the high-pressure plunger pump. His pump was also the first to have a variable plunger stroke as a means of increasing or reducing the amount of fuel to be injected.

Hans Grade, engine manufacturer in Magdeburg, Germany, began fuel-injection experiments about 1905. A Grade airplane from 1909 had a two-stroke engine with fuel

Injector patented by Edward Gardner in 1913 had fuel delivery from the side (h) into a pressure chamber, from which an overflow line (i) led back to the pump. The precise centering of needle movement was its major feature.

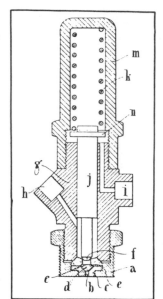

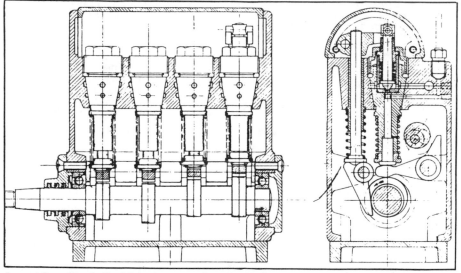

Layout for a Bosch diesel-injection pump for a four-cylinder engine from 1923 shows basic cam-and-plunger action for building up hydraulic pressure. A simplified device evolved after Bosch purchased the Acro patents in 1927.

injection, the pressurization of the fuel to be injected being provided by the charge-air precompression in the crankcase!

What these pioneer fuel-injection systems accomplished for the aircraft engine industry, however, was soon countered by the leading carburetor manufacturers. They developed new carburetors that did not mind being turned on the side or upside down; and the concept of fuel injection was robbed of a major incentive.

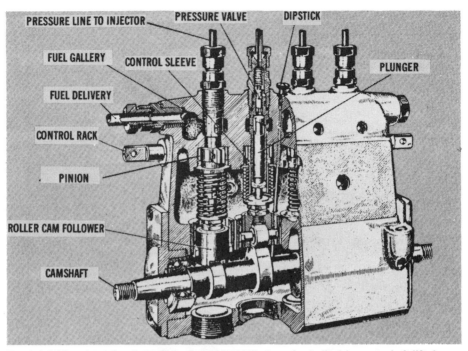

PRESSURE LINE TO INJECTOR   PRESSURE VALVE   DIPSTICK

FUEL GALLERY   CONTROL SLEEVE   PLUNGER

FUEL DELIVERY

CONTROL RACK

PINION

ROLLER CAM FOLLOWER

CAMSHAFT

Bosch injection pump for diesel engine, made in 1927. A mechanically driven camshaft lifted a set of plungers via roller-type followers. The plungers entered a fuel gallery and trapped a certain amount, which was then pushed into the pressure line to the injector. Fuel volume was varied by rotating the plungers by a rack-and-pinion arrangement. This four-plunger pump became the basis for Bosch and Daimler-Benz experiments with gasoline injection.

A British patent issued to Edward Gardner in 1913 covers a hydraulically operated injector nozzle with ground-in piston. It was remarkably advanced, having a taper-seat for the needle, so that it would always be centered on the orifice. Gardner obviously realized the importance of this point for exact control over injection timing and a clean-cut end of injection.

Robert Bosch of Stuttgart undertook a study of fuel injection in 1912, but it did not lead to practical results. It was not until the diesel engine had been invented and began to appear in trucks (Benz and MAN in 1923) that the demand for a high-pressure fuel-injection pump became a serious objective. Bosch made many attempts, and did have a measure of success.

But the real solution came from Switzerland, in the form of the Acro injection pump invented and patented by Franz Lang. It was announced in 1926, and shortly afterward Bosch announced its takeover of Acro AG.

By 1930, Bosch had perfected the diesel engine injection system to the point of being ready for mass production. But now that the injection equipment was available, the engine manufacturers and automobile companies showed no interest in adapting it for gasoline injection.

With the political and military reawakening of Germany that preceded the Nazi takeover of power, the DVL (Deutsche Versuchsanstalt für Luftfahrt—German Aviation Test Establishment) had been formed in Berlin. And it was DVL that brought Bosch back into research on gasoline injection (acting under orders from the Ministry of Transport in Berlin). Thus began the era of the high-pressure gasoline-injection system, with nozzles spraying directly into the combustion chamber.

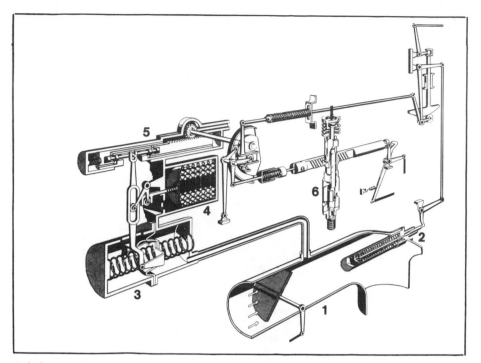

Aviation-gasoline injection system for the supercharged Daimler-Benz DB-601 fighter plane engine. Direct injection was used, and the pump metered fuel to individual injectors in accordance with manifold (1) pressure and temperature (from sensor 2). Pressure signal was received by transducer (3) whose output was modified by the aneroid bellows (4) for altitude adjustment. An amplifier (5) passed the message on to the linkage that worked the control rack. Only one plunger (6) is shown.

As early as 1932 the DVL reported substantial power gains in single-cylinder four-stroke test engines by discarding the carburetor and adapting the standard diesel pump for gasoline fuel. Both pintle-type and hole-type nozzles were used with success.

This work was carried out under the guidance of Dr. Kurt Schnauffer. The next step of progress was the DVL's conversion of a six-cylinder aircraft engine to fuel injection. It showed power gains between ten and seventeen percent on the ground-level dynamometer.

After this, Mercedes-Benz was pushed into fuel-injection research—as a major supplier of aircraft engines and a prominent customer of Bosch. In the fall of 1934 Mercedes-Benz began tests of a single-cylinder unit with direct injection and a Bosch pump. At first, the standard diesel oil filter was used as a fuel filter, and there was no leakage stop. Special filters were developed, a leakage stop was added and the nozzles were changed from the pintle type, spraying fuel onto the piston crown at an angle, to a design with multihole nozzles.

Then the single cylinder was incorporated into a V-12 block designated DB 601 or 2,061 cubic inches (33.8 liters) displacement. This engine went into production in 1937 with a starting power of 1,200 hp. From that point on, fuel injection conquered the aircraft field.

It was also in connection with aircraft engines that port-type injection was regarded with renewed interest. An Italian engineer, Ottavio Fuscaldo, who formerly designed O.M. passenger cars and trucks, had been working on aircraft-engine research and de-

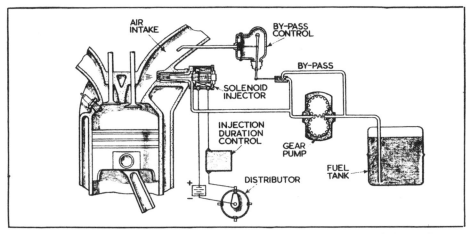

Fuscaldo fuel-injection system from 1940 worked with timed injection into each inlet port from a constant-pressure fuel line. The ignition distributor gave an engine-speed reading which was utilized to vary the duration of the injection.

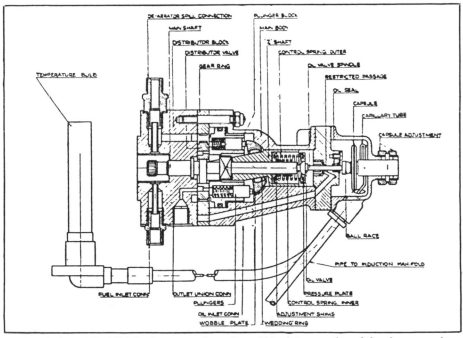

Sectioned view of the SU injection pump shows its wobble-plate operation of the plungers and fuel flow path through its hydraulic passages. Manifold pressure acted on a capsule with a capillary tube to affect metering.

velopment for Caproni from 1935 onward. It was out of dissatisfaction with even the best carburetors then available that he began to experiment with fuel injection.

The Fuscaldo system consisted of a gear-type pump that fed fuel under pressure through individual lines to each intake port. The injector nozzles carried precision-made

The principle of fuel injection for two-stroke engines is schematically explained by this sketch. Direct injection was needed to avoid spill of fresh fuel out the exhaust pipe.

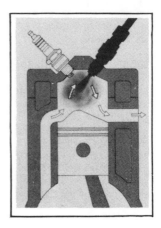

valves which were opened by electromagnets to spray fuel according to the needs of the engine. The system was used on an Alfa Romeo racing model that ran in the 1940 Mille Miglia, but was never applied to a production car.

On the allied side, SU Carburetter Company of Birmingham, England, developed a direct-fuel-injection system that came into use on Rolls-Royce Merlin aircraft engines toward the end of World War II.

Simonds Aerocessories negotiated a contract with SU for the American rights to the system, which was used by Continental for the Patton tank engine, an air-cooled 1,790-cubic-inch V-12 rated at 810 gross hp. The Patton tank came too late for seeing action in WW II, but was extensively used in the Korean conflict of 1950-53.

The postwar evolution of automotive fuel-injection systems started with the adoption of a Bosch system for two-stroke engines by two makers of small passenger cars in 1951: Goliath and Gutbrod.

That was the crowning achievement stemming from a study begun in 1931 by Dr. Kurt Schnauffer at DVL, when a two-stroke DKW engine was experimentally converted to direct fuel injection. Results were poor at first, but at last fuel injection succeeded in overcoming the worst drawbacks of the two-stroke: poor fuel economy, hard starting, smoke and noise. Since the two-stroke injection pump ran at crankshaft speed (instead of camshaft speed in a four-stroke engine) Bosch gained valuable experience with high-speed pumps. This was to be useful in later development work on four-stroke engines.

Bosch was able to apply technology from its aircraft-engine fuel-injection experience, with the main difference being the minuteness of the fuel quantities needed in these tiny two-cylinder engines. The volume of fuel needed for the Gutbrod engine would be no more than 1/1,500 of an ounce per delivery at 30 mph. Because of the necessity of using direct injection as a consequence of the two-stroke cycle, the timing had to be perfect and the metering highly accurate.

The injection pump was a two-plunger unit, the plungers being actuated by a cam and a lifter with a roller-bearing cam follower. The effective stroke of the plunger was varied by a pneumatic governor consisting of a diaphragm connected to the inlet manifold, behind the throttle plate.

The injector nozzles were inserted in the cylinder head adjacent to the spark plugs, with the injection timed to start at bottom dead center. The angle and shape of the spray were designed to draw maximum advantage from the natural turbulence in the two-stroke engine in order to assist the combustion process.

About 14,000 Gutbrod cars using this fuel-injected engine were built up to 1955, while only about 3,000 fuel-injected Goliaths had been turned out when the production was halted in 1957.

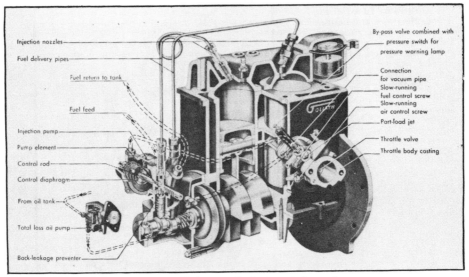

The labels on the figure are:

Injection nozzles
Fuel delivery pipes
Fuel return to tank
Fuel feed
Injection pump
Pump element
Control rod
Control diaphragm
From oil tank
Total loss oil pump
Back-leakage preventer

By-pass valve combined with pressure switch for pressure warning lamp
Connection for vacuum pipe
Slow-running fuel control screw
Slow-running air control screw
Part-load jet
Throttle valve
Throttle body casting

Bosche fuel injection for the 1951 Goliath two-cylinder two-stroke engine.

During the 1936-39 period, Mercedes-Benz had tested single cylinders from its Grand Prix engines equipped with fuel injection, with and without supercharging. No conclusive tests were reported before the outbreak of the war, causing the cancellation of the racing program and the closing of the passenger-car research department. Some fuel-injection tests were made with the 170 engine in 1946-47, but by then the injection pumps were obsolete.

Up to 1949, only direct injection with diesel-type nozzles was considered by the engineers in Stuttgart. That year, a racing car appeared at the Indianapolis Motor Speedway with a fuel-injected Offenhauser engine. The injection system was invented and developed by Stuart Hilborn aided by Bill Travers. It was *indirect* injection, a straightforward uncomplicated design, with a single throttle body at each intake port, feeding fuel continuously under pressure to spray nozzles inside them. This became known as constant-flow injection. The car did not finish the race, but by 1952 practically all entries had Hilborn's fuel injection.

It was later discovered that Ed Winfield, American carburetor inventor, had patented a port-type fuel-injection system as early as 1934.

At the beginning of 1952, Mercedes-Benz began work on a new injection system (with Bosch equipment) for the six-cylinder engine of the large 300 sedan. By the end of the year, a 300 SL prototype was running with a direct-injection engine, and tests showed a terrific power gain over the twin-carburetor version. It was developed for production, and became standard in the 300 SL gullwing coupe, beginning in 1954.

When Mercedes-Benz was planning its W-196 Grand Prix car of 1954-55, it was calculated that fuel injection would be necessary to reach the horsepower target. Here, cost was no object, and direct injection was preferred for this 10,000-rpm straight-eight 2.5-liter with desmodromic valve operation. The technical details of these and subsequent Bosch/Mercedes-Benz injection systems are given in a separate chapter.

About 1954 an American automotive engineer, Ben Parsons, who had been associated with the ill-fated Tucker venture, proposed a system of continuous fuel injection. Parsons attached a variable-speed centrifugal fuel pump, driven electrically, to the fuel tank. The electric motor was wired to a variable-output engine-driven generator, and thus 'knew' what rpm the engine was running at. That enabled it to adjust the fuel quantity

Constant-flow port injection was patented by Ed Winfield, famous developer and builder of racing-car carburetors, in 1934. A gear-type pressure pump sent fuel into a rail where a pressure regulator controlled fuel flow in accordance with throttle position and manifold vacuum.

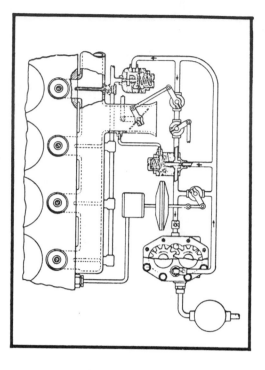

according to need. Injector nozzles mounted in the intake ports were designed to sense manifold pressure and meter the fuel quantities accordingly.

It was in 1955, after twenty years of research work, including continuous-flow systems, that American Bosch introduced a timed, mechanical system.

When Chevrolet was developing the V-8 Corvette, high-performance specialist Zora Arkus-Duntov began to look into ways of using fuel injection on it. Chevrolet engineers eventually developed a system that was put in production by Rochester Products Division.

It was a port-type system, overtly based on the Stuart Hilborn design, with clever modifications and overriding controls to give the necessary flexibility for a production-model sports car. Chevrolet made it optional for the 1957 Corvette.

Actual sales were delayed for months while Chevrolet engineers tried to sort out production problems. No more than 2,750 fuel-injected engines were installed in Chevrolet and Corvette cars during 1957. Ed Cole, then general manager of Chevrolet, said it would again be offered in 1958 but was "too costly and complicated for general use."

Oldsmobile and Pontiac had tried the Rochester system and rejected it. But Pontiac engineers made some modifications of their own, and used it as standard for the 1957 Bonneville. Service problems in the field caused both Pontiac and Chevrolet to go sour on fuel injection, and the Rochester system was no longer available when the 1959 models appeared.

From 1952 to 1961, all Offenhauser-powered Indianapolis-type racing cars used Stuart Hilborn fuel injection. European racing car constructors were well aware of this, and asked suppliers to produce competitive systems.

Lucas in Birmingham, England, developed a successful design for the 1956 Jaguar D-Type, which won at Le Mans. It led to a very expensive production version, and the only buyer was Maserati, for the 3500 GTI, beginning in 1961. Holley had purchased the American rights to the system in 1956, but found no market for it.

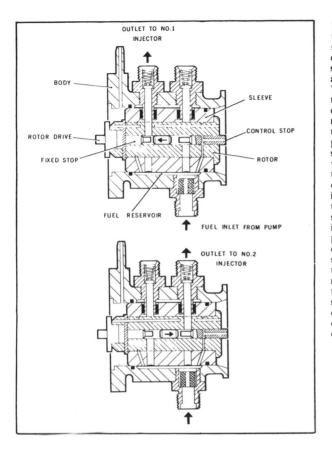

OUTLET TO NO.1 INJECTOR

BODY

SLEEVE

ROTOR DRIVE

CONTROL STOP

FIXED STOP

ROTOR

FUEL RESERVOIR

FUEL INLET FROM PUMP

OUTLET TO NO.2 INJECTOR

The Lucas fuel-injection system relied on a principle called shuttle metering for its basic operation. For the sake of simplicity, the drawings show a two-cylinder arrangement. The engine-driven rotor has two radial ports leading to a channel in its center. The rotor fits inside a sleeve containing fuel inlet and outlet ports, and its central channel serves as a track for a shuttle valve moving axially between two stops, one fixed and the other adjustable. Rotor ports index with the sleeve ports as the rotor turns within the sleeve, simultaneously reversing shuttle movement at the proper time. Fuel pressure drives the shuttle to the fixed stop, discharging to one injector. After 180 degrees further rotation, the shuttle arrives at the adjustable stop, fueling the second injector. The amount of fuel depends on the length of shuttle travel and the bore of the channel.

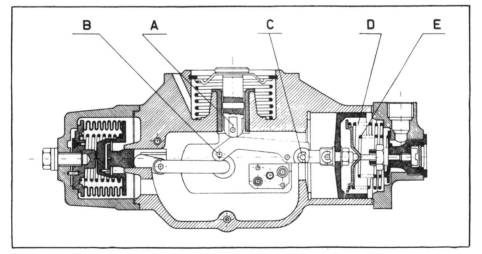

B   A   C   D   E

Lucas mixture-control mechanism is part mechanical, part hydropneumatic. Fuel-line pressure is brought to bear on the central spring-loaded diaphragm, and is relayed through roller A to follower link that carried rollers B and C. The central roller A acts on the pivoted follower which in turn influences the position of the fuel cam via roller B. Manifold vacuum is tapped into the space D-E to control the position of the diaphragm, which then moves the pivot point for follower B-C, exerting the major influence on the position of the shuttle control stop.

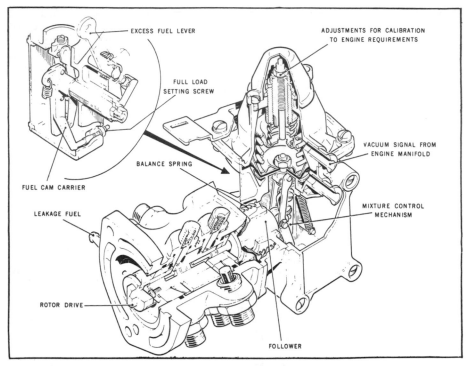

EXCESS FUEL LEVER

ADJUSTMENTS FOR CALIBRATION
TO ENGINE REQUIREMENTS

FULL LOAD
SETTING SCREW

VACUUM SIGNAL FROM
ENGINE MANIFOLD

BALANCE SPRING

MIXTURE CONTROL
MECHANISM

FUEL CAM CARRIER

LEAKAGE FUEL

ROTOR DRIVE

FOLLOWER

Lucas metering distributor and control unit, for a multicylinder engine. The outlet unions for
the pipes to the injectors incorporated nonreturn valves. These unions were screwed into the
body and passed through the sleeve ports in rubber sealing rings. The body was an aluminum
casting, and the rotor was made of hardened steel. The mixture-control mechanism adjusted
the shuttle travel distance in accordance with manifold vacuum or direct cam operation from
the accelerator linkage.

It was a port-injection system with timed delivery. Fuel was pumped to a distributor
at 100 psi by an electric pump. The distributor metered the fuel in accordance with the
mass airflow as measured by manifold vacuum, and took care of the timing by a mechanically driven rotor with outlet ports arranged to feed fuel to each injector nozzle when the
valve opened. Air intake was controlled by a sliding throttle valve linked to the accelerator.

The twin-rotor fuel-distributor unit incorporated a mechanical mixture-control
device responding to vacuum in the headers. Fuel quantity delivered by the pump was
always in excess of the demand, the surplus being drained off, and fed back to the tank.

The metering/distributor rotors were driven at half crankshaft speed from an auxiliary shaft. They contained ports which, during rotation, indexed with ports in an outer
sleeve as well as hollow distributing shuttles moving back and forth inside the rotors. Shuttle
movement was effected by means of fuel-line pressure from the pump, and the port align-
ment was designed to connect the shuttle bore with the supply line and the delivery pipe
to each injector nozzle in the proper sequence.

Shuttle movement was limited by two stops, fixed at the inner end and adjustable
at the outer end. The adjustable stop was connected to the mixture-control device, which
consisted of a slave cylinder with a spring-loaded piston that moved toward top dead
center under high-vacuum conditions and toward bottom dead center under low-vacuum
conditions. A connecting rod from the piston actuated a control wedge that lined up
against two reciprocating lifters aligned with the distributing shuttles.

Bendix fuel-injection unit was used on the Ford V-8-powered car that won the Indianapolis 500 in 1970. The system has four main components: mass airflow sensing, regulator, fuel metering and throttle and the injector nozzle. The airflow sensor works inside the main throttle body, with a combination main and boost venturi. An air diaphragm, fuel diaphragm and control valve form the regulator system. The air diaphragm is vented to the venturi airflow signal and creates a force which is opposed by the fuel diaphragm, which is vented to the fuel differential (proportional to the air differential). The fuel control or jetting system includes an idle valve to control low-end part-throttle operation and a main jet to direct fuel flow throughout the engine's power range. The nozzle system consists of pressurizing valves which keep the two banks equalized on fuel flow to the individual nozzles.

The stroke of the shuttles was dictated by the position of the control wedge— shortest when the piston was at bottom dead center, and longest at top dead center. By this method, the mixture was enriched during high-load conditions.

Lucas continued to develop the system, and an improved system was used on the BRM Formula One racing car in 1961. Later versions have come into general use on Formula One engines. With rare exceptions, it is used on all Grand Prix racing cars built since 1966.

At Indianapolis in the late 1960's, the supremacy of the Stuart Hilborn injection system was challenged. The challenge came from Bendix in the form of a mechanical port-type system developed for aircraft engines. This RS-11 system was used on the turbocharged Hawk that Mario Andretti drove to victory at Indy in 1969. The following year, Al Unser's Ford-powered Indy winner also had Bendix fuel injection.

This system was mounted upstream of the turbocharger and consisted of four main components: the mass airflow meter, the regulator, the fuel-control and throttle, and the nozzles. The airflow meter consisted of a throttle body with a combined main and boost venturi, giving an air signal proportional to mass airflow.

Bendix fuel injection on Ford-powered Indy car, 1970. The injection unit is mounted on the air-inlet side of the turbocharger. Mass airflow measurement proved to be a good approach to holding a fixed air/fuel ratio for all accelerator positions. It was fully satisfactory in racing cars, but did not offer the versatility needed for production cars.

The regulator positioned a fuel-control valve in accordance with the amount of air being admitted, by means of an air diaphragm that compared the mass airflow signal with the opposing force from a fuel diaphragm that was vented to the fuel-pressure differential across the jetting system.

This jetting system was made up of an idle valve to provide fuel for idle and part-load operation and a main jet for fuel flow throughout the power range. Air-bleed-type nozzles were inserted in the intake headers, vented to turbocharger outlet pressure.

By 1971 this system had been adopted by thirty-two out of the thirty-three cars qualified to start at the Speedway, and is still popular there. This brings us up-to-date on the racing front, but the developments just reviewed pale into insignificance compared with the progress that was being made in systems aimed at the mass-produced car.

The following examples, however, must be regarded not as prologues to future types of fuel injection but as the final attempts to devise new, improved systems with obsolescent technology.

No sooner had Chevrolet announced the Rochester injection system in 1957, than Borg-Warner Corporation began experimental work on a similar system. This project was assigned to the Marvel-Schebler Division and resulted in the development of a single-plunger injection pump, the distribution being arranged by rotation of the plunger to index with the discharge ports to each injector line in turn. The pump gave timed injection to nozzles installed in the inlet valve ports. Nozzles were of the spring-loaded pintle type, working at a pressure of about 200 psi. The pump carried a control assembly whose main input was manifold absolute pressure.

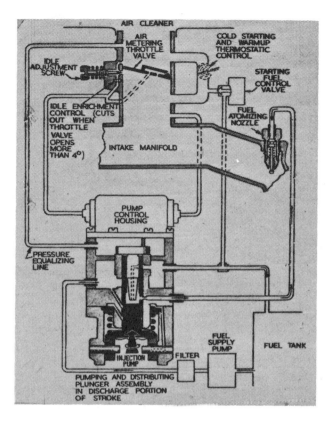

Schematic of Borg-Warner fuel-injection system, showing how all pumping action and metering functions are contained in one place: the injection pump.

The system was intensively developed in the 1960-66 period, but Borg-Warner could not sell it in Detroit. When electronic fuel-injection systems began to appear, Borg-Warner withdrew from the field.

American Bosch (originally a branch of Robert Bosch of Stuttgart, but independent since 1917) proposed a vaguely similar system in 1957, and Ford was interested, but dropped it when GM and Chrysler stopped offering fuel injection for production cars.

Like the Borg-Warner system, American Bosch used a single-plunger pump, driven from the engine camshaft. Face cams on the spring-loaded plunger indexed with a metering sleeve in a sliding fit around the plunger, which had a fixed-length stroke. The number of cams corresponded to the number of cylinders in the engine, and the position of the metering sleeve determined the spill point, thus regulating the amount of fuel to be injected.

The face cams operated on fixed rollers carried in the drive housing, producing reciprocating motion in the plunger. The plunger had a single discharge port, and rotation of the plunger lined up in sequence with the outlets to the injector nozzles. The nozzles were installed in the intake ports, and injection timing was automatically assured by plunger rotation. Fuel metering was accomplished by a control unit, mounted on the pump, with a linkage to the main throttle body, thus responding to manifold pressure. Commands from this unit were transmitted to the metering sleeve, which then moved axially along the plunger to blank off the spill ports as required.

A port-injection system with a simple, timed plunger-type metering pump was under development by TRW (Thompson-Ramo-Wooldridge) between 1959 and 1962, but it was never released.

Friedrich Deckel of Munich, long established as a manufacturer of diesel injection equipment for stationary engines and agricultural equipment, developed an automotive

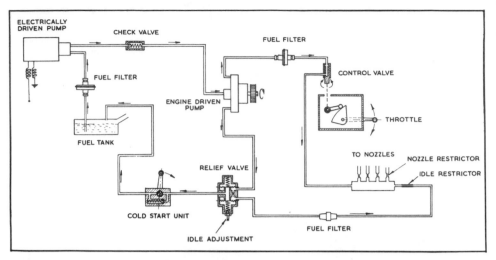

Schematic for the Tecalemit-Jackson system. In the engine-driven pump, where the fuel entered through the end cap and passed through a built-in filter, excess fuel was returned to the tank via a relief valve. The pump action relies on a toothed rotor and a discontinuous spiral groove in the housing cover. The rotor builds up pressure in the groove, at whose end the outlet is located. The pressurized fuel then flows to a control valve, mechanically linked to the accelerator, before arriving at the distributor which has individual outlets for each injector.

fuel-injection system in 1960. The Deckel company had been a subsidiary of Kugelfischer (Georg Schäfer and Company) since 1955, which explains why the system became known as Kugelfischer and not Deckel.

It was a port-type injection system with timed delivery. The pump contained a cam-actuated plunger for each injector, with fuel metering assured by an intricate regulator-cam arrangement. Air intake was controlled by a single butterfly throttle. Tests were made with Porsche racing engines, and Peugeot became interested. After thorough testing, Peugeot standardized Kugelfischer injection on the 404 sports models in 1962, and Lancia adopted it for some Flavia types in 1965. The fuel-injection branch of Kugelfischer was taken over by Bosch in 1974.

Continuous injection was a main feature of the Tecalemit-Jackson system, which appeared in its original form in 1964. It was tested by Ford, Vauxhall, Lotus, Jaguar and Aston Martin, but did not find any acceptance for production at that time.

The system was mainly mechanical, with electronic switches and port-mounted injectors. An electric pump fed fuel to an engine-driven diaphragm pump which increased the pressure and delivered fuel to the control unit, in which a metering valve and a sleeve regulated the amount of fuel on the basis of engine speed, manifold pressure, air density and various minor parameters.

At the heart of the Tecalemit-Jackson system was a ring-main in which fuel circulated continuously at pressures of up to 90 psi, variable according to speed and load.

Branch pipes extending from the ring-main carried the fuel to the injector nozzles, and excess fuel was drained off and returned to the tank. The inlet manifold was fitted with two throttle valves—one connected to the accelerator linkage, and the other serving as an upstream air valve to maintain constant vacuum in the space between them. It was also linked to the fuel metering valve via a cam arrangement, and a vacuum switch determined the opening duration for the injectors.

The Tecalemit-Jackson system was used on Mario Andretti's Ford-powered Honker, prepared by John Holman and Ralph Moody for the 1967 Can-Am racing program.

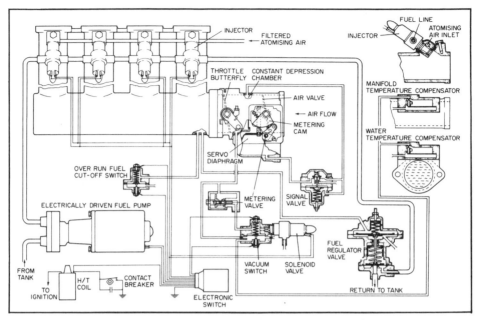

By 1974 the Tecalemit-Jackson system had evolved into this electropneumatic amalgamation with a single throttle body and airflow metering, pneumatically controlled fuel regulator valve and simple electronic switching. The constant depression chamber, framed by a throttle valve on the downstream side and an air valve on the upstream side, was kept at a constant vacuum, and had an orifice connected to a signal valve that reacted to any pressure fluctuations. Because of the signal valve's other connection to one side of a servo diaphragm that is part of the air-flow-measuring system, any drop in vacuum brought the servo into action to close the throttle plate. The engine-driven pump of the original system had been eliminated, and fuel was sent to the injector gallery at low pressure straight from the electric fuel pump. The injectors were electrically energized by electronic switching in accordance with signals from the vacuum switch and ignition contact breaker.

That year, a subsidiary company, Petrol Injection, Ltd., was formed at Plympton in Devon, England, to manufacture the system. Broadspeed and others used it for racing cars with Hillman and Ford engines, and it became optional for the Lotus-Cortina in 1967.

The Tecalemit-Jackson system was under evaluation by the motor industry as late as 1974, but by that time more advanced systems with lower production costs were being developed, rendering the Jackson patents worthless and forcing Tecalemit out of the fuel-injection market.

The 1969-model Alfa Romeo 1750 for the American market used a fuel-injection system developed by SPICA of Livorno, Italy, a Finmeccanica subsidiary that was well-known for its diesel injection pumps, oil and water pumps, transmission parts and for license-production of Lodge spark plugs, Allinquant shock absorbers and Burman steering gear.

SPICA fuel injection was later adapted to the 2000, the Montreal 2.5-liter V-8, the Alfetta and the Alfa Six, the latter two being current production models.

The SPICA system gave timed port injection with a plunger-type pump that was basically similar to those of Bosch and Kugelfischer. The plungers had constant stroke and a normal spill-port arrangement which routed excess fuel back to the pump reservoir. Fuel metering was dependent on manifold vacuum, throttle position, barometric pressure, idle setting and coolant temperature. Sensors worked in mechanical arrangements to affect the position of a regulator cone with a cam, free to move axially as well as rotationally, which was connected to the control rack for the injection pump.

SPICA fuel-injection system as used on the Alfa Romeo 1750. 1-Ignition key, 2-Fuel pump, 3-Fuel tank, 4-Fuel prefilter, 5-Main filter, 6-Injection pump, 7-Return valve, 8-Pressure control switch, 9-Pressure regulating valve, 10-Warning light.

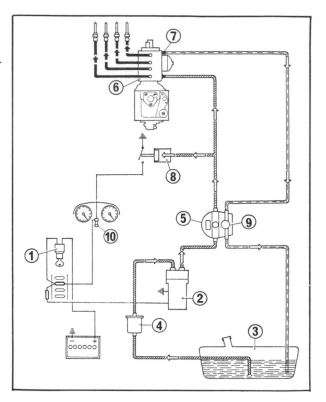

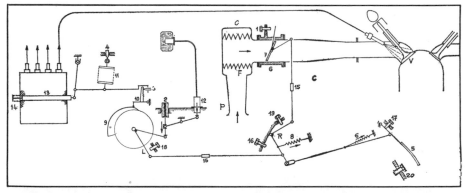

Schematic showing the SPICA installation on the Alfa Romeo 1750. The injector nozzle aimed the spray directly at the back of the valve head, and the accelerator pedal was linked directly to the injection pump. 1-Throttle-valve closing spring, 2-Idle-speed adjustment, 3-Mixture control at idle, 4-Mixture control under load, 5-Accelerator pedal, 6-Accelerator return spring, 7-Throttle plate, 8-Accelerator linkage return spring, 9-Injection pump camshaft, 10-Control lever, 11-Aneroid bellows (barometer), 12-Coolant temperature sensor, 13-Control rack, 14-Control rack return spring, 15-Throttle valve adjustment, 16-Full-throttle stop, 17-Idle-speed adjustment on pedal position, 18-Full-throttle stop on linkage, 19-Hot-engine throttle stop, 20-Full-throttle stop on pedal, C-induction system, F-air cleaner, G-throttle body, L-injection pump control lever, P-fresh air admission, R-bell-crank lever, V-inlet valve.

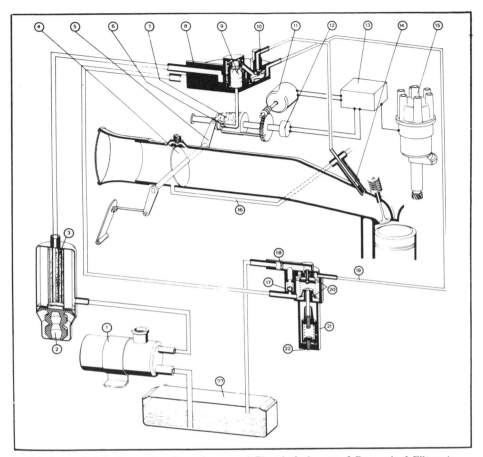

Zenith electronic injection system for racing cars. 1-Electric fuel pump, 2-Reservoir, 3-Filter, 4-Throttle valve, 5-Lever from cone follower, 6-Metering arm, 7-Cone, 8-Metering unit, 9-Metering piston, 10-Differential pressure valve, 11-Electric motor, 12-Potentiometer, 13-Electronic control unit, 14-Injector, 15-Ignition distributor, 16-Atomization air line, 17-Pressure control valve, 18-Pressure blocking valve, 19-Pressurized fuel line, 20-Differential pressure spring, 21-Pressure valve, 22-Pressure adjustment screw, 23-Fuel tank.

In the wake of the advances made by Bosch and Kugelfischer, Germany's biggest manufacturer of carburetors, Deutsche Solex, began a systematic search for new mixture-preparation methods in 1961. The initial work was done in collaboration with Professor Kurt Löhner and Günter Härtel of the Technical University of Braunschweig. Several patents were taken out in the 1961-1965 period.

The owner of Deutsche Solex, Alfred Pierburg, had also obtained the rights to Zenith and Stromberg carburetors, and renamed the firm Deutsche Vergaser Gesellschaft. In 1970 DVG set up its own research center at Neuss near Cologne, and the following year a separate fuel-injection department was created and placed under the direction of the India-born engineer, Asoke Chattopadhay.

This resulted in the development of a series of fuel-injection systems relying heavily on carburetor technology, presented under the Zenith trademark. Their technical details are explained in Chapter Eighteen. Suffice it to say, here, that apart from a few competition-car installations, such as Steinmetz-tuned Opels, and limited-production sports

The AE-Brico was one of the first electronic systems, being offered to the industry as early as 1966. It fell by the wayside, however; its one and only application being the Aston Martin DB6 Mark II of 1969. It offered timed and metered fuel delivery to individual injectors from a constant-pressure ring main.

cars such as the four- and six-cylinder BMW Alpina, the Zenith fuel-injection systems have not been accepted for use on production cars.

Today, Pierburg (as the DVG was renamed in honor of the founder) is collaborating with Bosch on the development of an electronically controlled carburetor.

In 1976 Holley offered the industry a new emission-control system which included a carburetor with electronic fuel metering. The system also comprised an oxygen sensor, electronic amplifier and catalytic converter. At one time Holley announced that one Detroit manufacturer would make it available on certain 1978 models, but the contract never materialized.

An electronically controlled carburetor was part of a new, advanced emission-control system that GM offered on about 1,500 cars for sale in California in the winter of 1977-78. It was made available first on four-cylinder Pontiac Sunbirds and V-6 Buick Skyhawks.

At the start of the 1981 model year, it was standardized on all domestic GM cars under the name CCC (Computer Command Control). The control unit activates a solenoid that operates an on/off needle valve that meters fuel to the main jet. The solenoid can open and close the valve up to ten times per second.

Inevitable or not, it was opportune that electronics should revolutionize fuel-injection technology, just as it had done in ignition systems and instrumentation. The principle of electronic fuel injection is very simple. The injectors are opened, not by the pressure of the fuel in the delivery lines, but by solenoids operated by an electronic control unit. Since the fuel has no resistance to overcome, other than insignificant friction losses, the pump pressure can be set at very low values, consistent with the limits of obtaining full atomization with the type of injectors used.

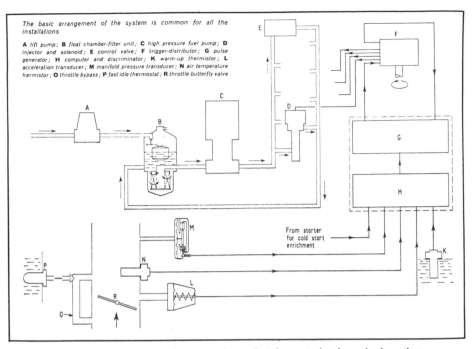

The basic arrangement of the system is common for all the installations

A lift pump; B float chamber-filter unit; C high pressure fuel pump; D injector and solenoid; E control valve; F trigger-distributor; G pulse generator; H computer and discriminator; K warm-up thermistor; L acceleration transducer; M manifold pressure transducer; N air temperature thermistor; O throttle bypass; P fast idle thermostat; R throttle butterfly valve

From starter for cold start enrichment

Schematic of the AE-Brico electronic system shows that the control unit received continuous readings on manifold pressure and manifold air temperature, as well as warnings of demand for sudden acceleration. The trigger-distributor varied the amount of fuel sent to each injector by variations in the duration of solenoid-valve opening. The electric feed pump delivered fuel to a float chamber combined with a filter, from where the fuel was drawn on demand from an electrically driven gear-type pump that raised the line pressure, limited to 25 psi by a control valve inserted in the ring main circuit. Despite its early appearance, the AE-Brico system's electronics were quite compact. The computer, discriminator, pulse generator and manifold-pressure transducer were contained in a common unit with aluminum base and heat sink, no more than ten inches long, 4.5 inches wide and 2.25 inches deep.

The amount of fuel to be injected is calculated by the control unit on the basis of information fed into it regarding the engine's operating conditions such as manifold pressure, accelerator enrichment, cold-start requirements, idling conditions, ambient temperatures and barometric pressure. The systems work with constant pressure, variable injection time or continuous flow. Compared with mechanical injection systems, electronic fuel injection has an impressive set of advantages—fewer moving parts, no need for ultra-precise machining standards, quieter operation, lower power loss (and low electrical requirement), no need for special pump drives, no critical fuel filtration requirements, no surge or pulsations in the fuel line and, finally, the clincher for many car makers, lower cost.

Studies of electronic control for fuel injection began when Robert W. Sutton, a Bendix engineer, tackled the problem in 1951. An experimental system was installed on a 1953 Buick V-8, which was used as a demonstrator to the auto industry.

Named Electrojector, the first Bendix system was used by American Motors Corporation as an option for the 1957 Rambler Rebel. AMC was never able to get the bugs out of the system, nor did the company get a flood of orders. In fact, no cars equipped with the Electrojector were delivered to the public.

In 1958, Chrysler began offering the Electrojector for the Chrysler 300-D and the De Soto Adventurer. A few hundred such cars were built during 1958 and 1959. The system was largely satisfactory, but needed further development. It was not free of service prob-

An early French candidate for a role in the electronic revolution was the Monpetit Sopromi system. A mechanical gear-type fuel pump pressurized the fuel and fed it to a fuel rail via a pressure regulator limiting the pressure to a constant 220-250 psi, with a return line. The electronic control unit received signals with information on atmospheric pressure, manifold vacuum, coolant temperature and engine speed. Solenoid valves integrated with the injectors gave timed injection with metered amounts by staying open for longer or shorter duration. 1-Fuel from tank, 2-Fuel pump, 3-Injector, 4-Pressure regulator, 5-Return line, 6-Control unit, 7-Atmospheric pressure sensor, 8-Manifold vacuum sensor, 9-Engine speed sensor, 10-Coolant temperature sensor.

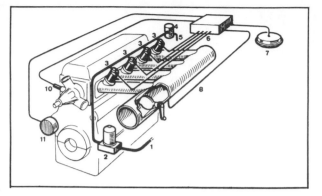

lems, for instance. Interference from outside sources could upset the electronic circuitry, and even the injector nozzles had some reliability problems.

By 1965 Bendix had a perfected system ready, and Bosch signed an agreement for access to the Bendix patents. That resulted in Volkswagen's use of Bosch electronic injection (later known as D-Jetronic) in 1968.

Cars that used the D-Jetronic in 1969 included Opel Admiral, Citroen DS 21, Volvo P-1800 E, Saab 99 E, Mercedes-Benz 250 CE plus all Mercedes-Benz models powered by the 3.5- and 4.5-liter V-8 engines.

They were followed by the Lancia Flavia in 1970 and Renault 17 in 1971. Volvo and Saab have since switched to K-Jetronic (without electronic control), and Mercedes-Benz uses the K-Jetronic on its fuel-injected four- and six-cylinder models. The Bosch K-Jetronic was first used by Porsche for the 911 T in 1973.

Bosch engineers began development on the K-Jetronic in 1967. The background to this project was a search for a simplified device to replace the high-pressure plunger-type injection pump. The main objectives were to eliminate the mechanical drive and to facilitate the installation with regard to space requirement so that it would be easily adaptable to a wide range of different cars.

These systems are described in detail in their respective chapters. Fuel-injection developments since 1970 cannot really be called history, but belong in the realm of current events. Evolution continues but it is logical to end the historical review with the start of the electronics revolution and the phasing out of gasoline injection systems using diesel-type plunger pumps.

5

Bosch Mechanical Systems for Mercedes-Benz Cars

Mercedes-Benz still offers a mechanical system, the K-Jetronic, for many of its models, but this chapter deals with the first, second and third generations of Bosch injection equipment for Mercedes-Benz cars—those preceding the K-Jetronic by many years and relying on more primitive methods.

As we have seen, in Chapter Four, when Bosch and Mercedes-Benz engineers began to tackle the fuel-injection problem, they relied heavily on diesel-engine technology and components. That was still the case for the World War II aircraft engines, and nothing had happened to set their thinking off in new directions when a new series of experiments was undertaken in 1947.

Both the 300 SL and the M-196 engines were created in a period when the use of direct injection was unquestioned in Stuttgart.

The Mercedes-Benz 300 SL was a high-performance sports car powered by a six-cylinder three-liter (183-cubic-inch) overhead-camshaft engine, derived from that of the 1951-model 300 sedan.

Each cylinder had a fuel nozzle inserted through the top of the cylinder block with its spray aimed directly into the combustion chamber under a pressure of about 1,100 psi.

The main fuel pump was mechanical, and was mounted on the injection pump, drawing fuel directly from the tank (via a pre-filter). In addition, an electrical fuel pump mounted outside the tank fed fuel under low pressure to a T-jointed return valve inserted in the main fuel line. From the mechanical fuel pump, the fuel was pushed through a fine filter and then fed into the injection pump.

The injection pump resembled the diesel type in its general layout and construction. The pump had a camshaft and six plungers, plunger movement being effected by the rotation of the camshaft. All the cams had the same contour, but they were offset relative to each other around the camshaft periphery so as to coincide with the proper firing order.

The plungers provided the pumping action that sent fuel to the injectors. When the cam lifted the plunger, the fuel held in the chamber above the plunger was forced out into the delivery line.

Fuel metering was accomplished by a spill-port, bleeding off excess fuel from each plunger when opened by a control sleeve. The sleeve had a number of slots and underwent partial rotation as directed by a rack connected to the accelerator linkage.

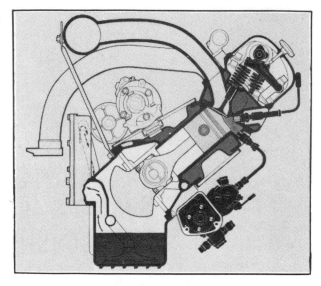

Mercedes-Benz 300 SL engine cross-section, showing injector positioning.

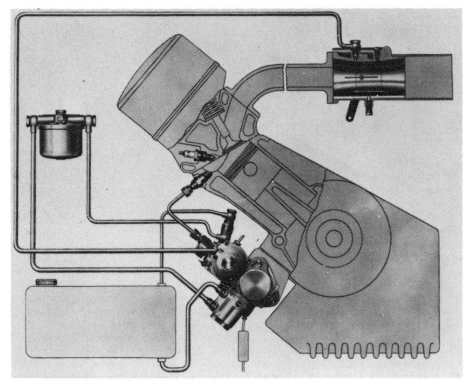

Bosch injection system for Mercedes-Benz 300 SL. Spark plugs were mounted in the cylinder head, and injectors in the side of the block.

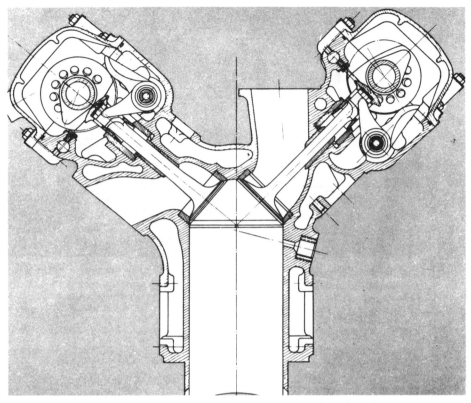

In the M-196 engine, the injector was inserted lower down in the block side, with the nozzle aimed against the center of the combustion chamber.

The 1954-55 300 SL had a Bosch PES 6 KL 70/320 R2 injection pump. An improved version, PES 6 KL 70/320 R3, was adopted in 1956.

The control rod for the injection pump was linked to a diaphragm placed between atmospheric pressure and manifold vacuum, thereby obtaining a measure of mass airflow. Cold-start enrichment was assured manually, from a button on the instrument panel.

Instead of a normal intake manifold, the engine was fitted with three pairs of ram pipes of arcuate shape, connecting the bazooka-like plenum chamber with the port flanges on the cylinder head. The ram pipes had a length of seventeen inches, which gave maximum ram effect between 5000 and 6000 rpm.

The Mercedes-Benz M-196 was an in-line eight-cylinder engine with dual overhead camshafts. It was used in 2.5-liter form in the W-196 Grand Prix car of 1954-55, and in three-liter form in the 300 SLR sports prototype of 1955. The M-196 engine layout permitted the use of long and straight induction pipes, giving considerable ram effect.

The injection pump was made by Bosch. It was an eight-plunger design, geardriven from a central power take-off on the engine.

The M-196 nozzles were of the pintle-valve type with a single delivery hole. Fuel pressure unseated a spring-loaded one-way valve at the right moment, and no return line for leakage was needed.

The injector nozzles were carried in holes drilled in the cylinder barrels. Extensive experiments were necessary to determine the optimum position of the jet itself and the

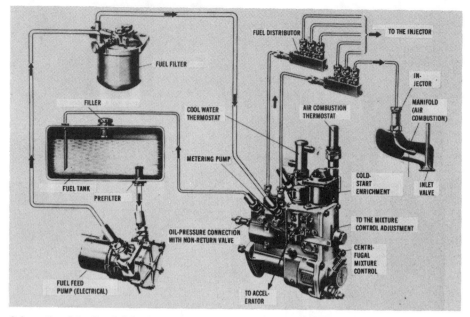

Schematic of the Bosch injection system for six-cylinder engines, with a two-plunger pump and triple-outlet fuel distributors, as used on the Mercedes-Benz 220 SE.

shape of the spray. Finally, the nozzles were placed on the inlet side of the block at an upward angle, aiming the spray directly against the exhaust valves.

On the supply side, a low-pressure pump brought fuel from the tank to a filter, and from there to the injection pump. A certain quantity of fuel, after passing a blowoff valve, returned to the tank. The purpose of this blowoff valve was to scavenge the inlet chamber on the injection pump and prevent vapor lock. Fuel delivery was accurately metered, pressurized at 1,500 psi, and timed to coincide with the compression stroke.

The fuel metering was governed according to the vacuum behind a throttle plate in a venturi at the plenum chamber entry, with part-load corrections made automatically in accordance with differential pressure across the throttle plate.

There was little freedom with regard to injection timing, since the pistons shrouded the nozzles at top-dead-center position. That left no more than 120 degrees of crankshaft rotation available for effective injection, which corresponds to only 0.002 second at 10,000 rpm.

With the arrival of the 300-d in 1957, Mercedes-Benz introduced timed port injection on a production car. It was no sports model, but a big and heavy limousine, using the six-cylinder three-liter overhead-camshaft engine from the 1951-model 300 sedan. This system was incredibly complex by modern standards and, therefore, so expensive that it could only be used on extremely high-priced cars.

The car was called 300-d (where the d just happened to be next in the sequence, following the 300-c, and had nothing to do with D for Diesel).

Injecting into the ports meant that the pump could run with much lower pressures (and consume less power) while making less noise.

The 300-d system gave metered and timed injection through six individual nozzles, mounted in separate runners for each port, and discharging in parallel with the airflow into the ports.

A plunger-type injection pump delivered fuel for each cylinder at 100 psi from an individual plunger, whose stroke was variable in accordance with throttle-linkage

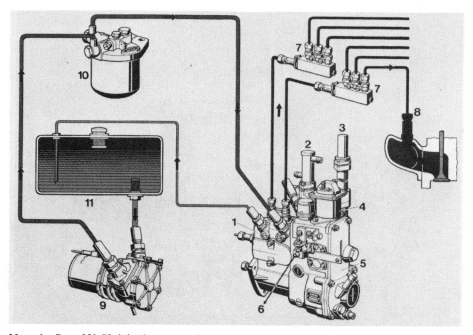

Mercedes-Benz 250 SL injection system featured a two-plunger pump with twin fuel distributors, providing out-of-phase injection timing for two-thirds of the firing impulses. 1-Return line pump, 2-Cooling water thermostat, 3-Charge air thermostat, 4-Solenoid for start enrichment, 5-Centrifugal governor, 6-Accelerator linkage, 7-Fuel distributors, 8-Injectors, 9-Electric fuel pump, 10-Fuel filter, 11-Tank.

position, and thereby provided fuel quantities in keeping with the requirements of the moment.

The control system consisted of a venturi to measure airflow, a fuel-feed pump, a magnetic enrichment device coming into action when the coolant temperature dropped below 40° C, a diaphragm to vary fuel flow in proportion to mass airflow at any given rpm, a thermostat providing a progressive leaning-out of the mixture with rising ambient temperature and an anemometer to reduce fuel flow in high-altitude operation.

The system also included a separate electrical fuel pump, intended to ensure rapid hot-starting by eliminating vapor lock effects. The pump was started automatically when the ignition was turned on and the oil temperature was above 100° C. In addition, it could be operated manually by a separate switch, to permit raw fuel to circulate throughout the system before starting the engine, or to purge the circuit of vapor locks after a hot-soak.

The 300 SE, introduced in September 1961, had a new aluminum-block engine, but continued using the fuel-injection system of the 300-d. The following year, Mercedes-Benz applied fuel injection to the lower-priced 220 model, for which a simplified system was developed. Instead of the six-plunger pump, it had a smaller and lighter two-plunger pump, providing semicontinuous and untimed delivery to all six nozzles through two distributor boxes, one for the front three and another for the rear three cylinders.

The two-plunger ZEB-series injection pump was then adapted for the 300 SE and applied to the 230 SL in 1963.

In 1965 Mercedes-Benz reverted to six-plunger pumps for the new 250 SE and 250 SL, ostensibly with the aim of obtaining surer, quicker starts under difficult conditions. The same six-plunger injection pump was then used for the 280 SE and 280 SL in 1968.

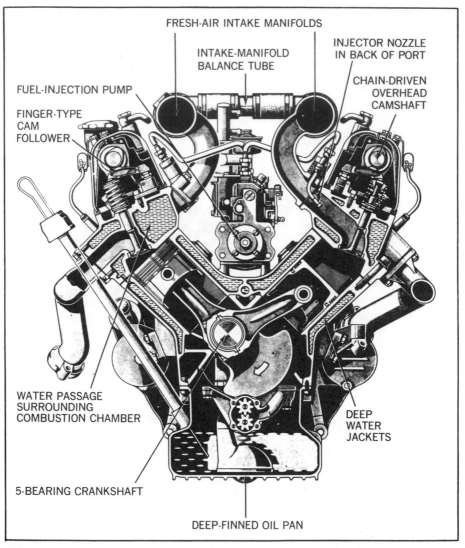

FRESH-AIR INTAKE MANIFOLDS

INTAKE-MANIFOLD
BALANCE TUBE

INJECTOR NOZZLE
IN BACK OF PORT

FUEL-INJECTION PUMP

CHAIN-DRIVEN
OVERHEAD
CAMSHAFT

FINGER-TYPE
CAM
FOLLOWER

WATER PASSAGE
SURROUNDING
COMBUSTION CHAMBER

DEEP
WATER
JACKETS

5-BEARING CRANKSHAFT

DEEP-FINNED OIL PAN

Cross-section of the Mercedes-Benz 600 engine shows the injection pump mounting between the banks of the V-8, with short pipes to port-type injectors inserted some distance back from the valves. Crossflow heads were adopted, with exhaust pipes on the outside of the V.

For the 250 SE engine, Mercedes-Benz went to a new type of nozzle mounting, inserted into a hole drilled in the cylinder head and aimed at the back of the valve head at an angle of about thirty degrees to the valve stem. This layout replaced the earlier nozzle type, which had been mounted on a manifold runner and included a long extension into the middle of the port area.

The mechanical governor of the 250 SE is also worthy of note. The aim was to obtain better coordination among all the various operational data affecting the metering system. The injection pump control rack was linked to a roller that rested on a spherical cam. This cam was part of the governor assembly.

The accelerator connection to the spherical cam was made so that changes in throttle opening rotated the cam about its axis. Axial movement of the spherical cam was controlled by a centrifugal governor. This governor filled the function of an engine-speed sensor, and was a high-precision unit equipped with sensors for coolant temperature and altitude. Air/fuel ratios were, therefore, automatically adjusted for cold-starting, warmup, cruising, coasting, sudden acceleration and changes in air density.

The roller moved up or down according to both axial and rotary motion of the spherical cam. As a result, the cam position dictated the movement of the plunger control rack.

Each surface point on the cam was a reference point for an exact combination of engine load, engine speed, water temperature and altitude (air density). Engine load signals were delivered to the cam by a connection to the throttle linkage. Thus, it was the throttle opening and not the actual volume of air entering the plenum chamber that determined the air/fuel ratio.

Injection was timed to the intake stroke, but injection starting point and duration varied with engine speed. At 1000 rpm injection began two degrees BTC, and the nozzle stopped injecting at thirty-six degrees ATC. At 4000 rpm, the nozzle began to inject fourteen degrees ATC and injection continued until ninety-eight degrees ATC.

The V-8 engine for the Mercedes-Benz 600 of 1964 had a new eight-plunger injection pump with individual lines to nozzles mounted in the port areas of the intake manifold, using similar controls and corresponding delivery timing.

Gradually the discovery was made by both Bosch and Mercedes-Benz that injection timing was quite unimportant with port-type injection systems.

There are obvious reasons for this, starting with the frequency of delivery. Even when the engine is running as slowly as 1000 rpm, each inlet valve opens four times in less than one second. At 5000 rpm, it is kicked open twenty-one times within a single second.

Inside the cylinder, it is almost impossible to distinguish one pulse from another, and the overall result comes to resemble continuous injection.

In 1968 the new 250 E coupe began using the Bosch electronic injection system, marking the beginning of a new era.

Kugelfischer Fuel Injection

In comparison with the contemporary Bosch injection system, the Kugelfischer was perhaps less accurate. Its merit lay mainly in its ingeniously simple system of mechanical regulation.

The fuel was drawn from the tank by an electric pump and sent through a fine filter before entering the feed pump. The output line from the feed pump led to a large paper-element filter and then to the injection pump, which contained a gauze filter at the entry port.

The pump housing was a two-piece aluminum casting that contained all the main elements—a camshaft with a plunger for each cylinder. An extension of the camshaft chamber contained the gears and bearings of the control mechanism as well as the ball bearing supporting one end of the camshaft. Fuel inlet valves were mounted on the lower face of the casing, with the outlets on top.

The amount of fuel that was injected per plunger stroke was regulated by the clever device of limiting the return stroke of the plunger. This was done by a mechanical stop attached to a vertical-pivoted lever whose position regulated the fuel metering.

This lever played a key role in the system. At one end, it was pivoted on an eccentrically mounted pin, while its other end made contact with a short pushrod. The eccentric pin could be manually adjusted to deliver extra fuel for cold-starting. The pushrod, which was actuated by a cam, had the delicate job of automatically regulating the amount of fuel in accordance with engine speed and load.

The stop on the lever was located near the middle and, consequently, its position could be changed by movement at either end, so that both the pushrod and the eccentric pin could influence the length of the return stroke.

The cam that controlled the pushrod had a three-dimensional profile, its spindle being geared to a magnetic core in a steel cylinder which was geared to the same shaft that turned the injection-pump camshaft.

Cylinder rotation around the core would apply magnetic drag, proportional to the speed, to the core, which felt it as a torque force. A gear train transmitted this torque to the regulator-cam profile.

Partial rotation of the cam imparted movement to the pushrod. In addition, the cam was arranged to move axially, which also translated movement to the pushrod, in response to commands from a simple rod-and-lever connection to the accelerator linkage.

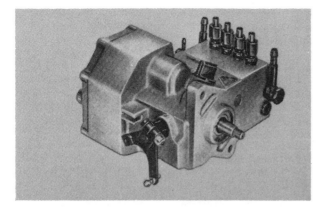

Kugelfischer injection pump for a four-cylinder engine.

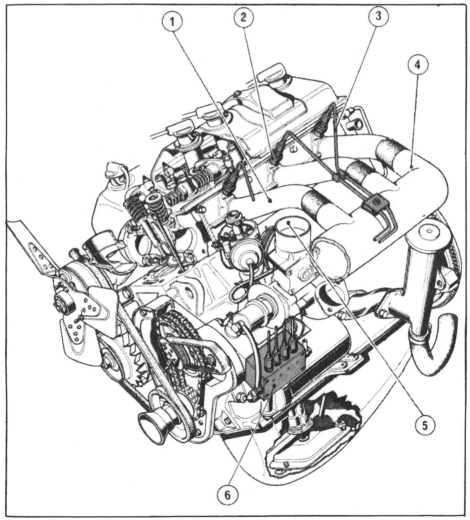

Kugelfischer installation on a Peugeot 404. 1-Intake runner, 2-Injector, 3-High pressure fuel line, 4-Manifold, 5-Throttle body, 6-Injection pump.

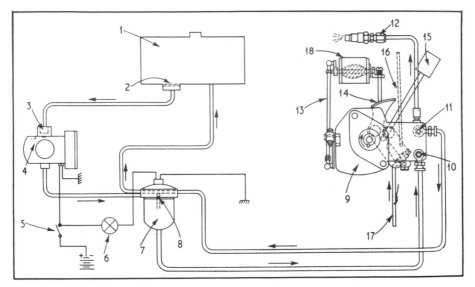

Kugelfischer system schematic. 1-Fuel tank, 2-Coarse filter, 3-Thimble-type filter, 4-Fuel-feed pump, 5-Ignition switch, 6-Warning light, 7-Main filter, 8-Air bleed, 9-Injection pump, 10-Fuel entry, 11-Excess fuel bleed-off, 12-Injectors, 13-Accelerator linkage, 14-Fast-idle cam, 15-Thermostatic control element, 16-Enriched-mixture control cable, 17-Oil pressure line, 18-Throttle body.

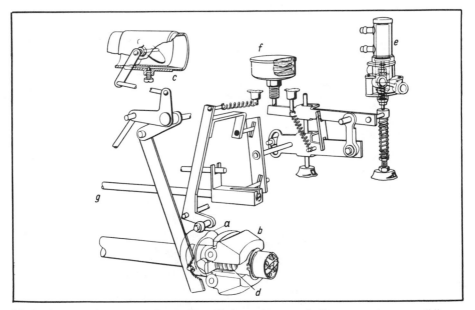

Mechanical metering controls for the Kugelfischer system were built up around a cone, sliding axially and turning on its axis. a-Cone, b-Centrifugal rpm-sensor, c-Throttle valve, d-Adjustment springs, e-Coolant temperature sensor, f-Aneroid for altitude compensation, g-Control rod.

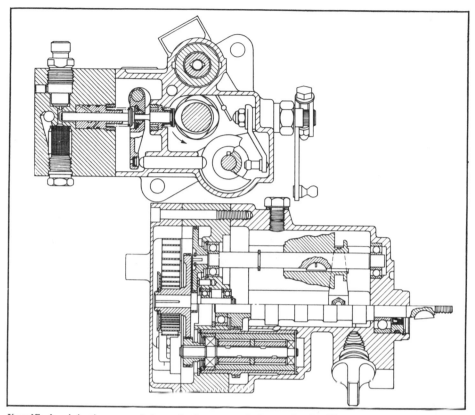

Kugelfischer injection pump belongs in the cam-and-plunger family. The gear on the camshaft drives the speed-sensing device. On its return stroke, plunger movement is stopped by the regulator lever, and a tappet return spring is used to keep the follower in permanent contact with the cam track.

In addition to the manual cold-start button, the control lever also carried a cam-actuated stop, which set up a fast-idle mode during the warmup period. Further, the lever that actuated the eccentric pin was connected to a thermostat, sensing the coolant temperature, which weakened the setting progressively as the engine approached normal operating temperature.

The cam was not fixed to the spindle, but merely keyed to it with a Woodruff key, which imparted rotational motion to the cam. Axial travel of the cam was assured by a peg on the end of the regulator lever from the accelerator linkage, the peg being registered with a groove machined around one end of the cam. A circlip in a groove on the spindle kept the cam from going off the spindle at one end. Its travel in the opposite direction was limited by abutment against the spindle bearing.

An injector nozzle was placed in each port. The main point of interest in the nozzle itself was the countercoiled spring, to prevent rotation of the needle.

Peugeot decided to install Kugelfischer fuel injection because of fuel economy rather than power gains. Tests showed that improvements of about fifteen percent could be attained in actual driving. At high speeds, the gains were even greater.

Naturally, Kugelfischer sold the system also on the strength of its advantages in enabling the motor engineers to raise power output. Accompanying the possibility of increasing compression beyond the knock limit of the carburetor version of the same engine, the accuracy of fuel metering assured a torque curve of improved shape and an absence of flat spots in operation.

Kugelfischer reported that in tests with two-stroke outboard motors, fuel savings up to forty percent had been realized.

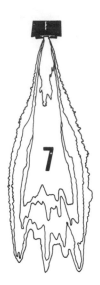

Rochester Fuel Injection

The widespread use of Hilborn-Travers fuel injection on Indianapolis-type racing cars no doubt influenced General Motors—and Chevrolet in particular—in its decision to develop and produce a fuel-injection system for passenger cars.

By 1952, when Indy cars, almost without exception, were equipped with fuel injection, GM was no longer a newcomer in this field.

Research work on fuel-injection systems had been undertaken about 1948 by the GM Engineering Staff in liaison with Allison Division in Indianapolis, which was interested in eliminating carburetors from its aircraft engines.

As had been the case with Bosch and Mercedes-Benz, GM opted for direct injection into the combustion chamber.

Conversion of a diesel-engine injection pump was undertaken, by adding metering controls. Within the narrow cruising-speed/full-power operating range of aircraft engines, the system did a satisfactory job of fuel distribution.

After thorough analysis of the experience, GM engineers concluded that by injecting into the intake ports, rather than the combustion chamber, the nozzle design could be greatly simplified. That would lower the cost and make it more attractive for automobiles.

The cost of a direct-injection system was far too high for general use in cars. GM engineers estimated its cost at eight times that of a carburetor system.

During the initial GM Engineering Staff tests, it was established that no substantial power loss would be suffered by going from direct to port-type injection. There would even be advantages: While the use of individual plungers for each cylinder gave sufficiently accurate metering for aircraft engines, it tended to give poor results in a car engine at idle and in city driving because of the erratic distribution pattern that was inherent with this system at low fuel-flow rates.

GM also looked at the Fuscaldo injection system, but felt that it could go further toward simplicity and obtain superior reliability.

No existing fuel-injection systems were considered satisfactory as a basis for further development. In consequence, GM engineers started to develop their own, guided by a few tests with single-cylinder laboratory engines.

These tests were quite thorough and proved definitively that direct injection did not offer any significant advantage over port-type injection. They also gave proof that timing of the injection was unimportant. In fact, some tests showed that there was a power

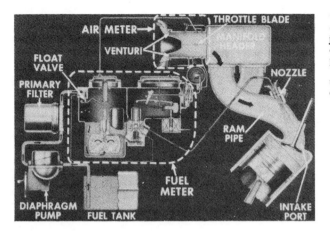

The Rochester system used by Chevrolet gave continuous injection into the ports with open nozzles in an atmospheric-pressure environment. The fuel line to the metering unit was pressurized by a diaphragm pump.

loss associated with timed injection, which also used more fuel than the rival setup with continuous injection.

Answers to the question of where to inject were clear. GM's tests showed that aiming the spray at the back of the valve head gave the most power, the lowest fuel consumption, the fastest warmup and the best acceleration response.

From this, the basic principles were settled. GM was to develop a port-type injection system with continuous flow, with open-orifice nozzles directing the fuel toward the valve heads.

At the outset, GM considered two different methods of air and fuel metering: speed density metering, as commonly used on aircraft, versus mass airflow metering. Speed-density metering was rejected after due deliberation, mainly because it requires means to accurately measure the engine's volumetric efficiency relative to rpm, which would be very difficult on engines working with ram-effect induction and, therefore, both complicated and costly.

Mass airflow metering, on the other hand, could be accomplished with a simple venturi. It was less than half as sensitive to variations in air density and temperature as the speed-density method, and fuel-density variations canceled a portion of the error.

The accelerator pedal controlled the volume of air admitted to the engine. Mass airflow was continuously metered, and a fuel meter discharged the right amount of fuel to form an air-fuel mixture of proper lean/rich properties for the prevailing conditions.

The incoming air was measured by an annular groove as it passed through the venturi, and a vacuum signal was sent to the main control diaphragm positioned above the spill plunger in the fuel meter. The venturi was made with an annular radial entry to obtain compact construction, low frictional losses and high venturi-metering signals from a relatively low pressure drop.

It is important to realize that the system depended on two vacuum sources: manifold vacuum and venturi vacuum. Venturi vacuum was created by a diffuser cone in the air intake to the air-metering side of the manifold. The vacuum in the venturi was so slight that it had to be measured in inches of water (while manifold vacuum was measured in inches of mercury on a scale 13.6 times as large).

The control diaphragm was designed to give an accurate response to air-pressure differentials of 0.01 inch of water, while also having the structural strength to withstand a thirteen-pound load, corresponding to a signal equivalent to fifty-five inches of water.

The highest pressure differential occurring in normal operation can be expressed as a ratio in excess of 3,000:1, and this differential affected the environment of both the diaphragm and the spill plunger.

The main control diaphragm was actuated by signals of venturi vacuum, and directed the action of the spill plunger.

A manifold-vacuum signal is transmitted to a variable-fulcrum control arm that is linked to the spill plunger. Depending on the strength of the vacuum signal, a diaphragm meters fuel by raising or lowering the control arm which in turn actuates the plunger.

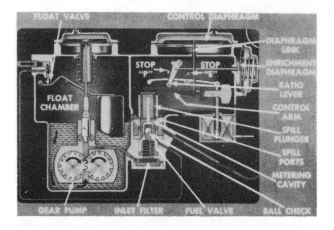

The smaller enrichment-control diaphragm was actuated by the stronger manifold vacuum. It directed the adjustment of the ratio lever, which gave automatic control of mixture enrichment.

The ratio lever was controlled by a spring-loaded diaphragm subjected to manifold vacuum. It moved to lean position at light load and rich position at full load, floating between the two during all transient conditions. Fuel was always present in the fuel meter, which contained a float chamber in which any vapors formed due to the temperature rise would be vented.

A conventional 6-psi diaphragm pump sucked the fuel from the tank and passed it to a ten-micron primary filter, and to the low-pressure section of the fuel meter. A small gear pump located below the float chamber raised the pressure and pushed the fuel through a second filter and a fuel valve to the metering cavity contained within the same housing, adjacent to the float chamber. A ball check in the fuel valve kept the pressure sufficiently high to eliminate vapor entry into the metering cavity.

The high-pressure pump was driven from the ignition distributor shaft at the same speed as the camshaft. Under normal driving conditions, fuel pressure generally stayed below 20 psi.

Some of the fuel delivered to the metering cavity was delivered directly to the injector nozzles, while excess fuel flowed through spill ports back to the float chamber. A spill plunger, linked to the air meter, regulated the amount of fuel spill.

It is immediately clear that this spill plunger was the most vital element in the fuel metering process. It worked by balancing the force exerted on the control diaphragm (by the static pressure drop in the venturi) against the force exerted on the spill plunger (by the nozzle metering pressure). The spill plunger was a variable bypass, constantly varying its position in response to signals from the two diaphragms.

The spill plunger was also designed to ensure that the fuel it allowed to drain back did so at normal float-bowl pressure. Any excess pressurization of the spill fuel would result in fuel injection into the float bowl. That, in turn, would cause loss of the fuel through the vents and serious malfunction of the system.

In normal operation, the air velocity, acting on the throttle blade, gave a signal of manifold depression, proportional to mass airflow. The venturi signal set up a force acting on the control diaphragm, which was transmitted to the spill plunger by means of a mechanical linkage.

Thus, increased airflow would give a stronger depression signal, putting a greater down-force on top of the spill plunger, reducing the volume of the spill flow, and causing an increase in the amount of fuel delivered to the injectors.

Theoretically, a constant air/fuel ratio was maintained, because the increase in fuel flow would be proportional to the rise in mass airflow.

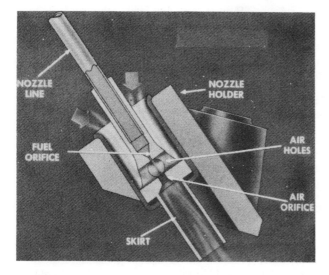

The injection nozzle is mounted in a plastic nozzle-holder and aims its spray through a skirt that directs it to the back of the valve head. The injector contains an air orifice for speedier atomization of the fuel immediately following discharge from the nozzle.

All the levers in the fuel metering system were counterbalanced so that their movement was not affected by their own weight. Their positions were determined exclusively by the forces exerted by the pressure-sensitive diaphragms.

Yet the system inherently offered some degree of enrichment at idle and near-idle speeds, because the weight of the plunger and the positioning of the nozzle below the spill level in the fuel meter worked together to admit more fuel under low-volume airflow conditions than would be possible with the basic venturi system alone.

This inherent enrichment was not sufficient, however, to assure smooth idling; so, more fuel had to be admitted from the idle orifice.

To change from power to economy, the ratio lever and its roller moved, thereby displacing the fulcrum point between the two opposing forces. The ratio lever was mounted on the same shaft as the enrichment lever, and both were actuated by the spring-biased enrichment diaphragm.

To accelerate from idle-speed, manifold vacuum intensified the venturi signal, and the off-idle orifice began to supply extra fuel. A restriction in the bleed passage regulated the amount of fuel drained to the off-idle orifice.

Because of instantaneous response from the fuel meter, no acceleration pump was needed. When the throttle was opened, air began to rush into the manifold in amounts exceeding the stabilized air requirement, which resulted in a correspondingly stronger venturi signal.

Since the parts that made up the linkage to the fuel meter were very light and operated with minute travel, their inertia was negligible. This meant that the inrush of air would cause an immediate rise in fuel pressure, and the injectors would receive an additional shot of fuel proportional to the increase in mass airflow.

During deceleration with a closed throttle, the amount of fuel vaporization that took place in the manifold was negligible. Additionally, a coasting shutoff valve was added on top of the float chamber to avoid supplying excess fuel that would show up in the exhaust gas in the form of unburned hydrocarbons.

Under normal conditions, the coasting shutoff valve was kept closed by a spring. The valve was also linked to a diaphragm subjected to manifold vacuum. During coasting, the rise in manifold pressure vacuum would exert a force on the diaphragm. When this force exceeded the load on the spring, the valve was pulled off its seat and allowed the fuel pump to discharge directly back into the float chamber. To prevent shutoff during other conditions of high manifold pressure (such as when revving the engine free of load), the valve was connected to the accelerator linkage and could not open except on a closed throttle.

For cold-starting, a solenoid was used to open a direct fuel passage from the pump to the injector nozzles. That was done because, at cranking speed, it would take twenty to thirty seconds for the fuel pump to build up enough pressure to unseat the check valve. The solenoid went into action when the starter key was turned on, and was deenergized when cranking stopped.

Gradual reduction of the cold-start enrichment during warmup was assured by a separate diaphragm, affected by manifold vacuum. The thermostat piston was also subject to manifold vacuum.

Under low-temperature conditions, the piston kept the ball valve below it closed, which blocked the diaphragm in full-enrichment position. During warmup, the piston would allow the ball valve to open gradually. And when no further enrichment was needed, the full manifold vacuum came to act on the enrichment diaphragm, regulating the ratio-lever position, and making the piston seal off its own vacuum-connection.

The injector nozzles had an accurately calibrated open orifice for the fuel flow. A small air chamber in the nozzle body, below the orifice, admitted filtered air from the air cleaner at atmospheric pressure through four holes.

This air supply made sure that the fuel discharge from the nozzle always occurred at or very close to atmospheric pressure, regardless of fluctuations in manifold vacuum. As a result, the amount of fuel injected was not subject to variables other than the metering system pressure.

The throttle was closed during idle operation, and about one quarter of the required air was admitted through the individual nozzle air chambers. The rest of the idle air bypassed the throttle blade through a separate channel, whose opening was regulated by the idle-mixture screw. Therefore it was the idle-mixture screw which determined the strength of the venturi signal caused by bleeding-in manifold vacuum from the idle orifice.

Because of the long distance from the injector nozzle to the valve load, the injector body carried a tubular skirt which kept the spray within a parallel formation of 0.040-inch maximum width, which was also claimed to assist in the mixing of fuel and air. The size of the fuel orifice was determined on a basis of maximum and minimum fuel-pressure needs—from an eight-inch fuel head at idle and during cranking, to a maximum of 200 psi under wide-open throttle acceleration and at top speed.

Despite all precautions and safeguards, the Rochester injection system proved to lack reliability in service. Moreover, Chevrolet and Pontiac engineers found that they were getting more power with carburetors (dual four-barrel and triple two-barrel setups were developed in the late 1950's). By 1959, Rochester's fuel-injection system was no longer offered on new cars, remaining available only from the spare parts bin.

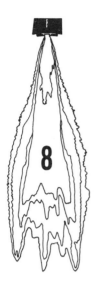

The Bendix Electrojector

The world's first electronic fuel-injection system for a gasoline-driven automobile engine was first described to an audience of automotive engineers at the annual meeting of the Society of Automotive Engineers in Detroit on January 15, 1957. Robert W. Sutton revealed that he had been working on the problem since late in 1952 in his laboratory at what was then the Eclipse Machine Division of Bendix Corporation in Lockport, New York.

He had filed for a patent on February 4, 1957, citing thirty-nine claims, which effectively constituted comprehensive blanket coverage for all forms of electronic fuel injection. The patent was granted on April 18, 1961 (number 2,980,090). Bendix then secured worldwide patent coverage for this system.

The goals were to produce a system that would be easily adaptable to existing engines, have a silhouette that would permit a further lowering of hood lines and have costs so low that it could eventually be used in mass production.

The basic invention was not a particular piece of hardware, but a complete system using fuel-injector nozzles working by means of solenoid-controlled valves, rather than fuel pressure against a spring load.

The introduction of electronic controls stemmed from dissatisfaction with mechanical means of fuel metering. "Our engineers were experimenting with various fuel management systems, but we had problems getting the mechanical device to perform the same way at all engine speeds," related Robert W. Sutton. "So I asked one of our engineers who was an amateur radio operator if there was any way to use electronics to help control the system. He felt there was."

Sutton reported that the engineer went out to a radio store and brought back some vacuum tubes and various other electrical components. Sutton and his colleagues did not know at the time that the use of electronic valves for delivering the fuel spray was not then a truly new idea. It was first tried in 1932 by an engineer named Kennedy, then working for the Atlas Imperial Diesel Engine Company.

He made a test installation on a six-cylinder low-compression, spark-ignition oil-burning marine engine. It did not have fully automatic controls and, of course, the transistor was not invented until 1948. Still, in 1934 Kennedy installed a smaller six-cylinder engine with the same system in a truck which was driven from Los Angeles to New York and back without reliability problems of any kind. The project died when the company folded, and was soon forgotten.

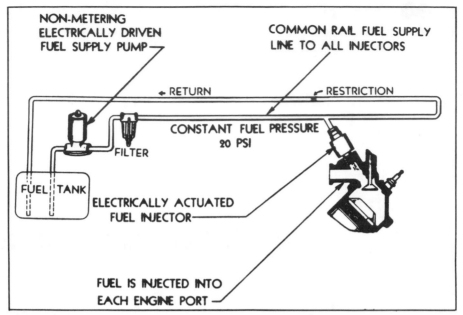

Bendix Electrojector system featured timed injection into the intake ports, with a 20-psi common rail fuel line.

The fuel injector included a valve operated by a solenoid according to electronic signals. Fuel entered at the top end and was discharged in spray form from the nozzle at the bottom.

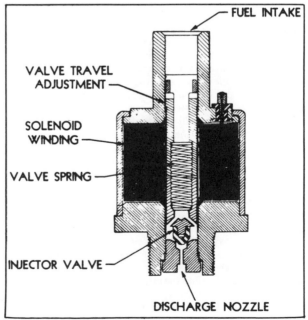

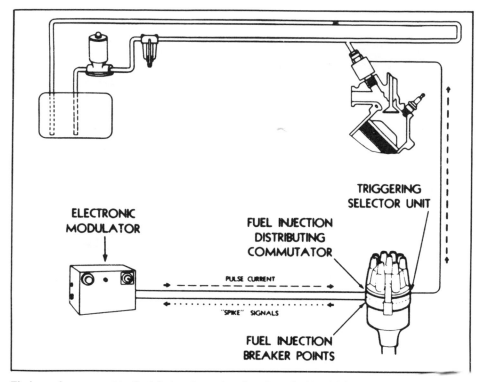

ELECTRONIC
MODULATOR

FUEL INJECTION
DISTRIBUTING
COMMUTATOR

TRIGGERING
SELECTOR UNIT

PULSE CURRENT

"SPIKE" SIGNALS

FUEL INJECTION
BREAKER POINTS

Timing pulses are sent to the injectors from the triggering selector unit (assembled as part of the normal ignition distributor) Opening duration is determined by the electronic modulator in accordance with speed and load.

The Eclipse engineers started from scratch, blissfully innovating in areas where experts had stayed out, from fear of running into too many difficulties. "The system we came up with was pretty primitive, but it showed promise," Sutton continued. "The problem was that none of us were electrical engineers, so we thought we had better consult electronics specialists from Radio Division to help us iron out some of our difficulties."

To prove that the device was workable, Sutton and two other engineers installed it on a 1953 Buick V-8 and drove it down to Towson, Maryland, with the whole electronic control unit sitting on the front floor.

"Once we got to the Radio Division, we discussed with their engineers what we had in mind," explained Sutton, "but they said there was no way to electronically meter fuel for an automobile engine. We gave them a ride in our car and afterwards they were enthused and agreed that it could possibly be done."

William L. Miron, vice president of Bendix, later recalled, "You can imagine the problems such a delicate system would have encountered in the harsh automotive environment. You can also imagine the cost of those components nearly a quarter of a century ago. Yet the basic system performed surprisingly well, if you were prepared to accept the fact that a slight problem with the electronics often resulted in an inoperative engine.

"And the cost? As one of our engineers observed not too facetiously at the time, 'We have a new electronic fuel-injection system that is almost as reliable as a fourteen-dollar carburetor; and only costs twenty times as much.' "

As Robert W. Sutton recalls it: "We had so many problems with the vacuum tubes that we had to switch to transistors, although at the time they were very expensive.

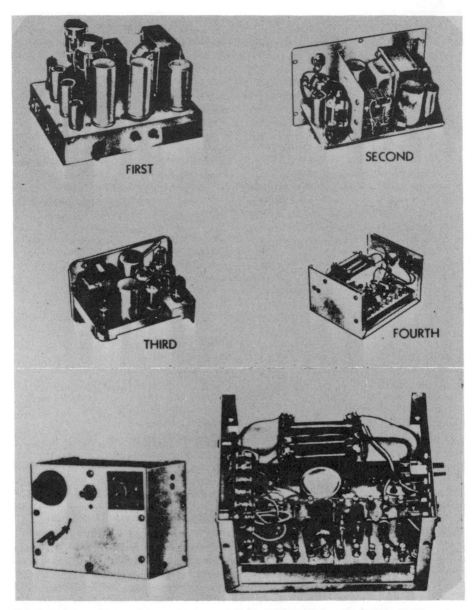

FIRST

SECOND

THIRD

FOURTH

Four generations of electronic modulator were built and tested before the first production unit was developed. Bendix began with vacuum tubes, which needed time for warmup before they could be brought into function. Size came down progressively. The final version is not shown to the same scale as its predecessors, but corresponds closely with the dimensions of the fourth-generation unit.

"For instance," Sutton went on, "when we used the vacuum tubes, it took thirty to forty seconds from the time we turned on the ignition for the tubes to warm up before the engine could be started. Also, sometimes when we drove under high-tension power lines the induced current would trigger the modulator to make the fuel injectors run freely and the engine would load up with fuel."

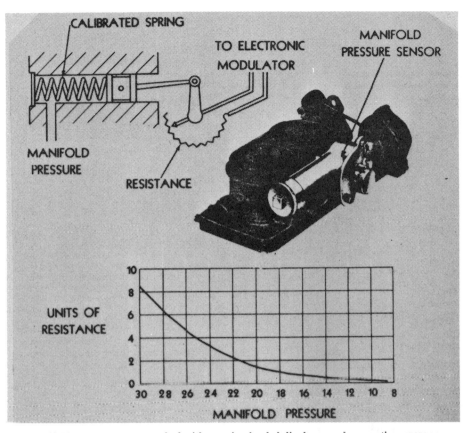

The manifold pressure sensor worked with a spring-loaded diaphragm whose motions were relayed by a crank mechanism that enabled a transducer to send quantified signals to the modulator.

Despite the problems, Sutton obtained corporate approval for continued research on a more intensive scale. The Research Division, the Radio Division, the Friez Division and the Scintilla Division of Bendix were contacted to contribute their expertise in particular areas toward the advance of the project.

Eventually, the Eclipse engineers developed an all-transistorized control unit that became the 'brain' for the Electrojector. The transistors themselves were produced at the Red Bank Division of Bendix. The unit required twelve transistors plus two silicon power transistors that cost about twelve to thirteen dollars each.

Several years later, William L. Miron analyzed the experience: "Our engineers couldn't develop, with the then-available electronics, suitable packaging to protect the system from the environment sufficiently to insure acceptable reliability. Nor could they develop an acceptable interface between the mechanical injection system and the electronic controls to assure the required levels of performance."

The basic principles behind the Electrojector were electronic control and electric actuation, with timed injection into the ports from a low-pressure (20 psi) common rail fuel system. It made use of a control system that responded to intake manifold pressure, engine rpm, ambient temperature and air pressure.

The choice of the common rail system stemmed from the basic premise that with individual lines, there were inertia effects at play in them, which upset the timing of fuel delivery.

Starting enrichment—and its tapering-off during warmup—was assured by a thermostatic control system, brought into action by the starter key. The throttle body carried a conventional fast-idle cam for warmup conditions.

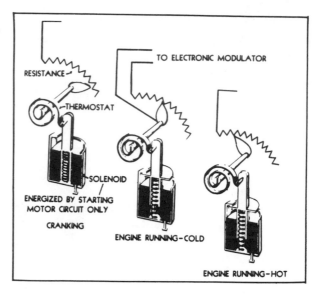

RESISTANCE

TO ELECTRONIC MODULATOR

THERMOSTAT

SOLENOID

ENERGIZED BY STARTING MOTOR CIRCUIT ONLY

CRANKING

ENGINE RUNNING-COLD

ENGINE RUNNING-HOT

Fuel was drawn from the tank by an electrically driven feed pump that maintained the fuel pressure to the fuel-injector valve, at 20 psi, plus or minus 0.5 psi. An ordinary fuel filter was inserted between the feed pump and the injector valves. A fine filter was not necessary, because the system had no close-fitting mechanically operating units.

The events of the operating cycle occurred in this sequence: By the cranking of the engine, the breaker point began to send out spike signals to the control unit in time with the opening of each intake valve. This triggered a multivibrator circuit in the control unit and set up a series of pulse currents tuned to a standard pulse width.

Airflow into the engine was controlled by a throttle valve of normal type but bigger dimensions. It gave a fairly accurate measure of mass airflow. A pressure sensor was mounted on the intake manifold, transmitting signals to the control unit about air density.

The various sensing units worked by adding a resistance to the basic circuit, thereby modulating the pulse width of the electrical impulse to be transmitted to each injector valve in turn. Pulse widths increased as resistance in the circuits increased, which caused the injector valve to be held off the seat for an extended time, permitting additional fuel delivery.

The injectors were connected to the fuel rails and carried solenoid valves which started and stopped the fuel delivery.

Fuel entered at the top of the injector valve, passing through the center core of the valve, and discharging through the nozzle at the lower end when the valve was off its seat (open).

Bendix reported that it took a major experimental effort to develop an injector valve that could operate at high speed, maintain calibration, run with low power demand and be manufactured at a reasonable price.

The solenoid valves were working in conjunction with an engine-driven commutator having segments shaped to vary the solenoid-energizing time in relation to engine characteristics and changes in operating conditions.

With regard to the spray formation, Bendix found that the best results were obtained when the fuel was directed at the back of the inlet valve head, so that a minimum of wetting of the manifold walls would occur.

Timing was obtained by adding a fuel-injection triggering selector unit and rotor to a standard ignition distributor. These elements were inserted as a sandwich between

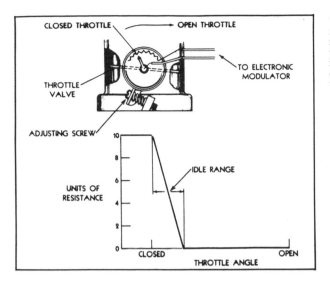

To give separate idle enrichment, a rheostat was connected to the throttle shaft, throwing a variable resistance into the control circuit whenever the throttle plate returned to closed position.

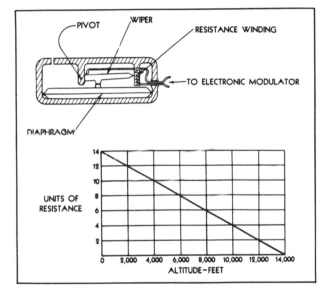

Compensation for high-altitude operation was accomplished by a device including an aneroid bellows with a variable-resistance unit that transmitted a signal to the modulator.

the base and the distributor cap. The triggering selector unit contained a set of breaker points and a commutator divided into sections corresponding to each injector valve.

The breaker points were actuated by the same cam that was used for the ignition system. By this method, it was assured that the fuel-injection breaker points made and broke contact, for each two revolutions of the crankshaft, as many times as the engine had cylinders.

Each time the fuel-injection breaker points made contact, a triggering impulse was transmitted to the electronic control unit. The modified signal was then returned to the selector element, and the impulse forwarded to the correct injector valve.

The primary function of the electronic control unit was to transform the spike signals received from the triggering-selector unit into an electrical impulse of a given standard width.

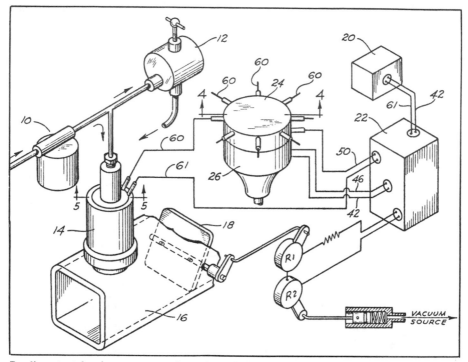

Bendix patent drawing serves as a clear schematic for the Electrojector with the triggering selector unit (26) in the center. The pump (10) pressurizes the fuel, but line pressure is limited by the regulator (12) before it reaches the injector nozzle (14).

It also had to coordinate engine-operating signals from sensing units located on different parts of the engine, and integrating these external signals into the standard pulse-width circuit—and thereby modifying it in response to operating conditions.

An amplifier boosted the pulse current to the power required for energizing the injector-valve solenoids. The amplified pulse was returned to the selector element and distributed to the correct injector in the same manner that a spark plug receives an ignition current. The solenoid lifted the injector valve off its seat and fuel delivery began in the port area, from where the airflow carried the mixture into the cylinder.

As we have seen, fuel delivery increased when the pulse width increased. Bendix pointed out that fuel delivery was not directly proportional to the pulse widths, but followed a characteristic curve, with the strongest rise in fuel quantity accompanying the initial growth in pulse width, tapering off under continued growth.

The standard pulse segments were modified in accordance with input data from other sensors or overriding manual controls. Other sensors, in the prototype installation, included the manifold vacuum sensor, the altitude compensator, the deceleration cutoff sensor, starting enrichment control, idle-mixture enrichment control and acceleration enrichment control.

The pulse width remained constant during wide-open-throttle acceleration, resulting in a fuel delivery rate proportional to engine rpm through the major portion of the speed range.

Bendix found that the fuel-delivery curve would begin to dip above 4000 rpm, but that in a typical engine of that time, this was matched by a slowing down in the growth of air consumption, canceling the risk of air/fuel ratio disorder.

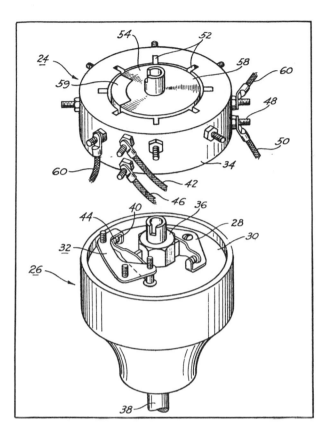

Modifications required to transform a regular-ignition distributor into a fuel-injection triggering selector unit proved relatively simple. The switch (32) is mounted on a stationary contact block (34) carrying a number of output brushes corresponding to the number of cylinders. Note that 50 is an input lead and 48 an input brush.

A return fuel line was part of the system, continually purging air or fuel vapor from the supply system. Tests with high under-hood temperatures and a shutdown period with empty fuel lines showed that these conditions did not lead to engine malfunctioning.

A separate idle-enrichment control was provided to assure smooth idling. It consisted of a rheostat connected to the throttle shaft, throwing a variable resistance into the control circuit when the throttle was closed.

Cold-start enrichment was accomplished by connecting a thermostat to a solenoid that could only be energized by the starter motor. The thermostat would position a variable resistance in the control circuit so that the pulse width would shrink with rising temperatures. For starting, the solenoid positioned the resistance to give increased pulse width during the cranking period. During warmup, a fast-idle setting was obtained by a conventional (carburetor-type) fast-idle cam and thermostat mechanism.

Bendix included provision for the insertion of a thermistor in the intake manifold to send signals about actual air temperature entering the engine. This signal would also be a resistance, transmitted back to the control unit to modify the pulse width and fuel delivery as required.

No acceleration pump was needed, since the manifold-vacuum sensor was arranged to provide signals for mixture enrichment when required for acceleration.

Rapid change in manifold vacuum would cause the breaker points to separate, thereby introducing an additional resistance into the circuit and widening the operating pulse of the signal to the injectors long enough to equalize the pressure on both sides of the diaphragm in the sensor. In case that proved to be insufficient, Bendix was also considering mechanical methods, with a connection to the accelerator linkage.

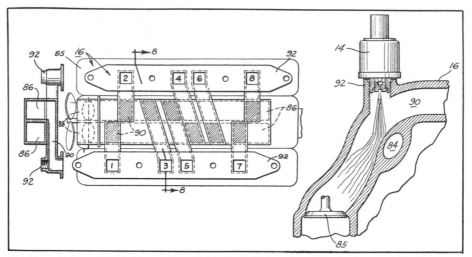

The idea of alternate fuel delivery to each bank of a V-8 engine, with simultaneous injection from four nozzles, was part of the original Bendix patents. The injector position revealed in the patent drawing may have been intended to mislead competitors, for there is no evidence that Bendix ever thought a spray directed against the manifold runner wall far back from the valve would be the most efficient method.

A fuel shutoff, going into action during deceleration with abnormally high mani-fold vacuum, was provided, mainly with the motive of making emission control easier. It would play a part also in avoiding fuel waste.

By adding an aneroid bellows with a variable-resistance wire, it became possible to provide the control unit with signals about air density, and thereby obtain automatic altitude compensation.

And yet, despite high levels of refinements, the Electrojector was not successful. Why not? Robert W. Sutton explained in a few words: "The main problem with the Elec-trojector was that electronic components were so expensive that the system was priced right out of the market. Also, people at that time were more concerned about horsepower than fuel economy and emission control."

Without original-equipment contracts, Bendix put the project on the back burner in 1960. William L. Miron recounts: "One of my first assignments when I came to Bendix in 1961 was to kill our electronic fuel-injection program. At that time, Bendix had already invested more than $1 million in a system that appeared to hold little promise as a viable product for the corporation because of the poor cost/benefit ratio and its apparent inabil-ity to survive in the harsh automotive environment."

Miron continued: "So we shelved the program and it wasn't dusted off until several years later when two major developments occurred: first, the demand for a more accurate method of metering fuel in order to reduce exhaust pollutants; and second, technological advances which made an electronic fuel-control system more economically feasible. These technical advances included the development of wire harnesses and connectors capable of transmitting electronic signals at the critical junctures, suitably rugged circuit designs and circuitry, mechanical mountings and interfaces that could survive the auto-motive environment—a host of interrelated problems had been resolved."

The renewed effort was to awaken Cadillac's interest and lead to Cadillac's use of Bendix electronic fuel injection on the 1976-model Seville.

Bendix had granted patent licenses to Bosch in 1966 and 1968 for electronic fuel-injection manufacturing rights in Germany and Brazil, and for world sales except for Canada and the United States.

In 1968 Bendix and Bosch also signed a mutual technical assistance and cross-license agreement. Then in 1969, Bendix and Bosch jointly negotiated license agreements with Nippondenso, Japan Electronic Control Systems Company, and two years later a similar contract with Joseph Lucas Industries. In 1974, the Bendix-Bosch cross-license and technical assistance agreement was broadened to include zirconium oxide exhaust-gas sensor technology.

9

Cadillac and the Bendix Analog Fuel-Injection System

Cadillac had never been tempted to use the Rochester continuous-flow mechanical fuel-injection system. Chief Engineer Carlton Rasmussen was convinced it could only hurt, and not enhance, Cadillac's reputation. And in the early days of emission controls, Cadillac had no incentive to change from the Quadrajet carburetor.

But after 1970, Cadillac began to suffer from the same driveability problems that affected most American engines at the time, with stumbling and stalling, jerky automatic-transmission shifts, cold- and hot-starting difficulties and running-on (dieseling). Fuel consumption rose sharply in the 1970-73 period, and it was in March 1973, that Cadillac started a serious test program with the latest Bendix electronic fuel-injection system.

Bendix had let the Electrojector lie dormant until 1967, when the engineering brains at Bendix saw an opportunity for its use arising through the pending exhaust emission control standards.

In the meantime, Bosch had developed its own electronic fuel-injection system (under Bendix patents) which went into production for the '68-model Volkswagen Type 3 (1600).

"We started to take a new look at the work we had done on our fuel-injection system because we felt that with its accurate fuel-control potential, it could contribute to lowering emissions," said John Campbell, then general manager of the Motor Components Division of Bendix.

"After looking at the fantastic developments that had been made during the decade in electronics, we decided to dust off our hardware and data and do additional development work to see exactly what contributions an electronic system could make to lowering emissions."

An engineering group was formed at the Elmira plant in 1968 to study the system's potential. "Our studies indicated that we did have potential for providing auto manufacturers with the tools they needed to help meet emission legislation, while still providing fuel economy, performance and good driveability," explained John Campbell.

In 1970 the group began moving its operations to the Detroit area to keep in closer contact with the automotive manufacturers, and in 1972, the Electronic Fuel Injection Division was established with headquarters in Troy.

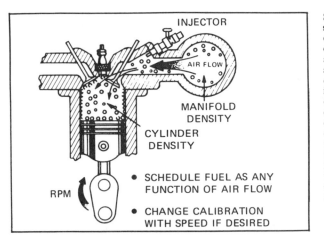

INJECTOR

AIR FLOW

MANIFOLD
DENSITY

CYLINDER
DENSITY

RPM

- SCHEDULE FUEL AS ANY
  FUNCTION OF AIR FLOW

- CHANGE CALIBRATION
  WITH SPEED IF DESIRED

Schematically, the speed-density control logic becomes clear. Fuel metering is based on mass airflow measurement as determined by engine speed and air density. For any given engine configuration, airflow is proportional to the product of cylinder air density and engine speed. Cylinder air density is equal to the product of manifold air density and the volumetric efficiency of the engine (its degree of cylinder filling). On the basis of these factors, fuel flow can be scheduled as any desired function of airflow.

Bendix electronic control unit from 1974. The experts knew it was only a starting point and were already planning two phases of adapting new technology, but this was what the company then proposed for production.

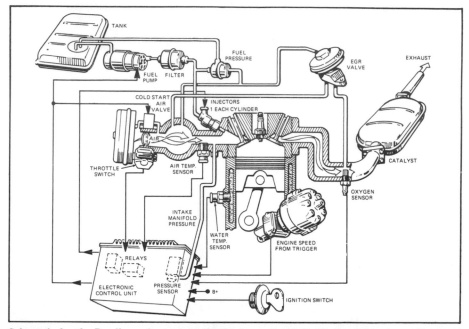

Schematic for the Bendix analog system of 1974. It convinced Cadillac's engineering staff that electronic fuel injection was now ripe for production in America.

Throttle body for V-8 engine. Its basic function is to adjust the volume of air entering the engine in response to the driver's commands. In addition to the throttle valves which are linked to the accelerator pedal, it includes systems for startup air control and idle-air bypass.

By that time, semiconductor technology and electronic-systems manufacturing techniques had ripened to the point of making the electronic fuel-injection system a realistic possibility.

William L. Miron outlined the commercial and technical reasons for the new situation: "With transistors, integrated circuits, thick film, conformal coating and so

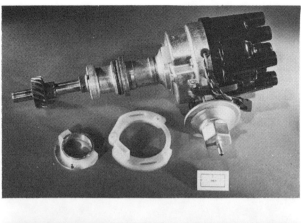

The earlier triggering selector was replaced by a simple speed sensor, feeding a continuous flow of data on engine speed and phasing to the control unit. These data serve to relate the frequency of injector operation and the time at which injection starts. The sensor shown here incorporates magnetic-reed switches which work without any rubbing or sliding contact.

Fuel-pressure regulator typically operates at a natural frequency of 400 Hz with a gain of 0.043 psi per gallon per hour flow over a flow-rate range of five to forty gallons per hour. In the Bendix system it was preset to maintain a constant pressure of 39 psi above the intake-manifold pressure. It bleeds off excess fuel (and pressure) via a flat-plate valve varying the bleed-flow orifice area in response to the prevailing balance between fuel pressure and a force representing manifold absolute pressure.

forth, the cost and complexity has diminished sharply, while the price and complexity of that fourteen-dollar carburetor has risen tremendously. With high-volume production, the cost differential between electronic fuel-injection and carburetion is virtually nil, when we consider the entire vehicle system."

The contract with Cadillac became possible, first, because GM's most prestigious car division was under pressure to improve driveability and fuel economy and, second, because Bendix was not offering just parts, but a complete system. In September 1975, it became standard equipment on the 1976-model Seville and optional for all full-size Cadillacs.

The importance of the second condition was underlined by Miron in a statement made as early as 1974: "Tied in very closely with the need to reduce costs is the absolute necessity that any electronic package designed for the automobile must be approached on a systems basis. It's economic suicide for any supplier to attempt to approach an electronic systems package on a component-by-component basis. It must be done as a total package, taking into consideration the total requirements of the vehicle."

Compared with the Electrojector, the new system was different in componentry but not in principle. "Its principles were always sound," a Cadillac engineer asserted. "It was mainly a matter of refining the functions for greater accuracy and reliability, and taking some of the cost out of the system."

Some of the biggest differences were to be seen inside the control unit, its components and circuitry. It is also noteworthy that fuel-line pressure was almost doubled, to 39 psi. With regard to timing, a new adaptation was required because Cadillac had

Both the air temperature sensor and the water temperature sensor have the same physical design as shown here. They are two terminal devices comprised of a coil of high-temperature-coefficient nickel wire sealed in an epoxy case and molded into a brass housing. The resistance of the wire varies as a function of temperature, sensor output being linear over the whole temperature range. The voltage drop across it is continuously monitored by the control unit.

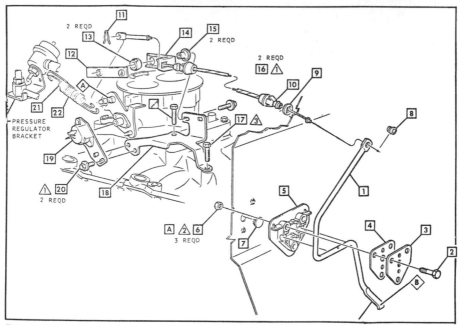

Despite the simplicity of its principles, the linkage to the Cadillac throttle body became a complicated assembly.

Fuel injectors for the 1975 Cadillac V-8 are solenoid-operated on-off valves with poppet-pintle nozzles. They deliver a narrow-angle spray cone aimed at the intake valve, minimizing manifold wall wetting. Typical injection duration times range from 2.5 to 9 milliseconds, with an opening time of 1.7 milliseconds and a closing time of 1.2 milliseconds.

Combined with the Delco high-energy ignition distributor, the speed sensor is mounted on the shaft below the head. It contains a pair of rotating magnets, whose rotational rate is picked up by reed switches in the cover plate and relayed as electronic signals to the control unit.

adopted the Delco electronic high-energy ignition system in 1975, and means had to be devised for enabling the breakerless distributor to serve as an rpm sensor, with a magnet assembly and a reed-switch assembly.

A pair of rotating magnets carried on the distributor shaft, interacting with a pair of reed switches, were encapsulated within the housing cover plate. When the magnets flew past the reed switches, they generated a make-and-break current, accurately counting the engine revolutions. Why two magnets and two switches? Because the injectors were divided into two groups which were separately timed.

Injection timing was modified so that the system's differences of principle from a continuous-injection system became blurred. The injectors of the four central cylinders were set to spray at the same time, during one revolution; the four cylinders on the front and rear ends received fresh fuel during the following revolution, all at one time.

The injectors were of the pintle-nozzle type, with solenoid-operated on/off valves. The injectors were inserted in the intake ports and aimed at the back of the valve heads. The amount of fuel only varied according to nozzle-opening duration, which was dictated by the electronic control unit.

The electronic control unit received a continuous flow of data from five sensors: the manifold pressure sensor, the throttle position sensor, the coolant temperature sensor, the ambient air temperature sensor and the engine rpm sensor.

The system was called 'analog' because it relied on an analog device for measuring mass airflow. One of the refinements over the original Electrojector consisted in indexing fuel-line pressure to manifold *absolute* pressure.

The manifold absolute-pressure sensor had a diaphragm that compared vacuum with barometric pressure. The analog sensor generated a voltage proportional to manifold vacuum, and thereby informed the control unit of variations also in speed and load.

This way, airflow rates could be determined within closely defined limits, since airflow is proportional to the product of cylinder air density and engine speed. Cylinder air density differs from manifold air density by taking the engine's volumetric efficiency into account.

The throttle body had the task of controlling the main airflow. This was done by a pair of conventional throttle valves linked to the accelerator pedal. The body contained an idle air channel, which had an adjusting screw for setting idle speed (with warm engine).

A fast-idle valve was installed on top of the throttle body and was arranged to allow supplementary air to enter, bypassing the throttles, during warmup. The valve had its own thermostat, which gradually closed the valve in accordance with both time and temperature.

Fuel was supplied from an electric booster pump submerged in the tank to the constant-displacement roller-vane-type main fuel pump positioned outside the tank, driven by a direct-current electric motor. The main pump had the task of maintaining a pressure of 39 psi throughout the system, with a pressure regulator referenced to intake manifold absolute pressure.

Cadillac's system was an unqualified success, being regarded as one of the most advanced production-car fuel-management systems then available anywhere in the world. And the role played by Bendix cannot be overestimated. An example from the production scheme will serve to illustrate how the supplier's involvement overlapped with that of the car company.

Bendix received shipments of intake manifolds from Cadillac at its Troy plant in Michigan; installed the injectors, fuel rail and throttle body; tested and calibrated the complete assembly; and returned it to Cadillac for mounting on an engine.

Meanwhile, Bendix engineers were making rapid progress toward what they thought of at the time as the 'ultimate' in fuel-management systems.

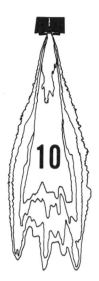

10

Cadillac and the Bendix Digital Fuel-Injection System

As the electronic revolution progressed and its promises for future cars were better understood, Bendix and Cadillac also worked together to simplify and improve the electronic-ignition system.

While Bendix was still tooling up to make the analog-control injection system for Cadillac, its engineers were well aware of several ways to improve it. Jerry Rivard, chief engineer of the Electronic Fuel Injection Division, pointed out in October 1974, that the 1975-model control unit was a hybrid, representing a mix of different technologies.

Standard control functions were handled by custom-made bipolar integrated circuits making use of emitter-coupled logic. Off-the-shelf bipolar integrated circuits and discrete elements were tried out for new functions then in the research stage. Experiments were also conducted with ceramic thick-film substrates to form active subcircuits and passive resistor networks.

According to Rivard, Bendix was frequently asked why digital circuits were not being used for the electronic control unit. The answer was that Bendix was working toward that exact goal. Patents were applied for in 1970, 1971 and 1973.

The programming of a digital microprocessor requires serial computation as opposed to the parallel computations that take place in analog computers. The problems that had to be overcome lay partly in developing sensors to supply analog signals from all operating parameters having a bearing on air-fuel mixture needs, and partly in the amount of computation that needed to be done in a minimum of time in order to convert analog data to digital information.

For a long time the sensors had shortcomings which made them something of a road block on the way to progress in this direction. About 1972 Bendix began testing miniature semiconductor strain-gauge pressure sensors. It also tried out, on cars with solid-state ignition systems, a system in which electronic signals from the breakerless distributor replaced the magnet-actuated reed switches.

Jerry Rivard revealed to an engineering audience in Detroit in February 1976, that he expected the then-current hybrid-analog system to be replaced by a new design based on digital metal-oxide-semiconductor technology.

This control-unit concept would permit the elimination of the previous pulse-forming network and gating by substituting a standard printed circuit containing all circuit elements and networks, housed in a small metal case.

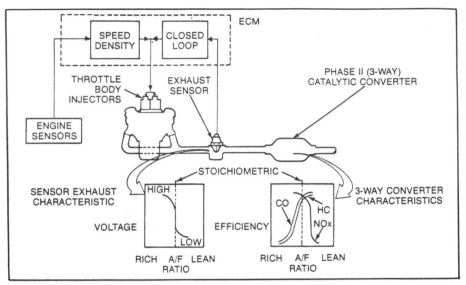

The Bendix digital fuel-injection system adopted for the 1981 Cadillac V-8 included throttle-body injectors, and was combined with a Lambda-Sond exhaust feedback device.

Manifold absolute pressure was chosen as the main criterion for fuel metering, because it reflects both speed and load variations. Signals from the vacuum sensor go through a linear variable differential transformer that feeds a flow of data to the control unit.

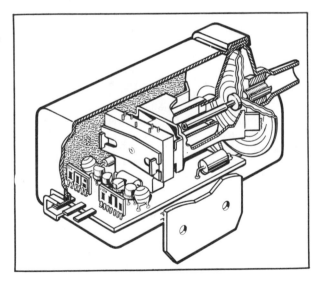

"As electronic-circuit technology continues to evolve," Rivard stated, "we are today evaluating two types of digital circuits. One utilizes metal-oxide-semiconductor large-scale integration in combination with read-only memories. The other utilizes metal-oxide semiconductors in large-scale integration, with a microprocessor. The microprocessor approach provides a true on-board computer which can time-share sensors and electronics to implement optimum control of numerous functions ranging from fuel management to dashboard displays."

Such a microprocessor would operate from algorithms and the non-volatile memories of metal-oxide semiconductors. Because of the high cost of such systems, important savings could be realized in combining other control functions, such as ignition, with

Bendix had been experimenting with sequential multipoint injection and proved its advantages over the two-group injection method, but Cadillac chose further simplification and went to throttle-body injectors.

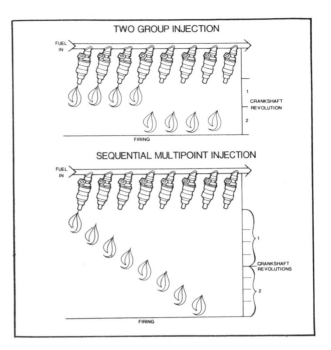

the fuel system, forming a single, integrated electronic system using the same control module.

This line of reasoning led Bendix to run two parallel research programs: one for developing the fuel-injection system itself; another for integrating spark timing, exhaust-gas recirculation, supplementary air injection and other emission-control systems with the fuel-injection control module.

A Bendix engineer, J. N. Reddy, pointed out that a closer look at engine requirements revealed that considerable benefits could be obtained by varying the duration of the pulses applied to the injector valves with increased frequency—ideally for each crankshaft revolution—and to effect these changes in pulse duration independently of intake manifold pressure.

That could only be achieved with digital control, and experimental technology was tried out on various combinations, as outlined earlier, by Jerry Rivard.

Once a basic approach had been agreed on, Bendix modified the standard fuel-injection system on a 1978-model 350-cubic-inch Cadillac V-8 to function with sequential multipoint injection instead of the two-group timed injection. The results were more than impressive, for the exhaust analysis showed a forty-three-percent improvement in carbon monoxide emissions, a twenty-eight-percent improvement in unburned hydrocarbons and a seventeen-percent improvement in oxides of nitrogen emissions.

This test marks the first application of digital engine control. Reaction from the Cadillac engineers was not slow in coming, and Bendix and Cadillac worked hand-in-hand to prepare a production system for 1979-model-year introduction.

Cadillac was worried, however, about the extra cost of sequentially timed port injection, and a compromise solution was reached. The eight injector nozzles were removed from the port areas and replaced by just two injector valves enclosed in a central throttle body, mounted like a carburetor on top of the intake manifold.

This was an important deviation from earlier principles, removing the mixture-formation stage from the inlet ports to an area far upstream, which was regarded as both 'good news' and 'bad news' at the same time. On one hand, the fuel was given much more

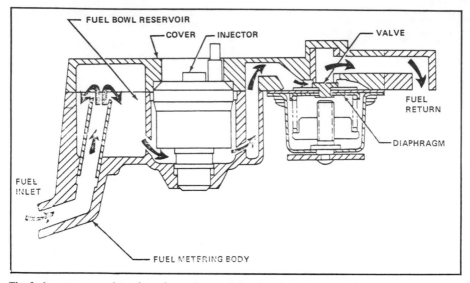

The fuel pressure regulator is an integral part of the throttle body. A relief valve operated by a diaphragm acts in response to atmospheric pressure to balance the fuel-line pressure. A constant pressure drop is maintained across the injectors.

A new principle was adopted for the throttle-body injectors: Fuel delivery was taken straight to the nozzle end, without going through the injector valve from the top.

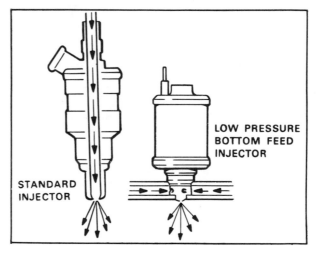

time to atomize and mix with the air, which worked in favor of a more homogeneous mixture capable of lean-burn operation. But, on the other hand, it introduced the risk of wetting down the manifold walls with raw fuel during periods of mixture enrichment.

Cost eventually turned out to be the deciding factor. The Bendix digital control system with central throttle body injection was first offered on the 1980 Cadillac Eldorado and Seville, and extended as an option for all Cadillacs (except diesel-powered models) for 1981.

Fuel is pumped from the tank and delivered to the throttle body after filtration. The pump is a twin-turbine type, electrically driven, built in unit with the tank float unit. Pump operation is controlled by the control module via a pump relay. It goes into action when the ignition is turned on.

Schematic view of the Cadillac injector. The injector valves are electronically actuated. A solenoid valve opens the ball valve by raising the plunger, as in port-type injectors. The fuel is metered into the throttle body upstream of the throttle blades.

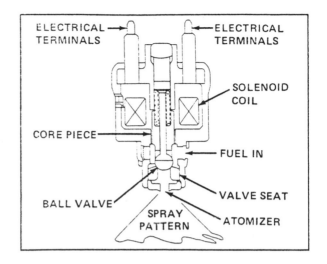

ELECTRICAL TERMINALS

ELECTRICAL TERMINALS

SOLENOID COIL

CORE PIECE

FUEL IN

BALL VALVE

VALVE SEAT

SPRAY PATTERN

ATOMIZER

The throttle body contains a pressure regulator which maintains fuel-line pressure at a nominal 10.5 psi, plus two fuel-injector valves metering fuel into the throttle body. They are positioned above the throttle plates, aiming the partly atomized fuel into the throttle bores. Injection timing, as well as the amount of fuel, is dictated by the control module.

Why bother with timing for injectors mounted so far away from the cylinders? The answer is not self-evident, but becomes clear when you look at the airflow system.

The incoming air passes through a filter and then enters the throttle body, where fuel is injected. A special distribution skirt is part of the lower throttle body, directly below each injector valve, to assure uniform distribution to all parts of the intake manifold—the timing being unrelated to any single cylinder but corresponding exactly to the average mixture requirement of all the cylinders.

Throttle valves control the airflow rate in response to accelerator-pedal movement, and idle speed is determined by the position of the throttle valves in combination with the idle-speed control.

The throttle body is a cast-aluminum housing with two throats and two throttle blades mounted on a common shaft. One end of the shaft is connected to the accelerator linkage, and the other end to the throttle sensor.

The manifold absolute pressure sensor, manifold air temperature sensor and the barometric pressure sensor give the principal inputs serving to measure the amount of air entering the engine.

An rpm signal was provided by the Delco High-Energy Ignition system. On the basis of these inputs, the electronic control module makes high-speed digital computations to determine the amount of fuel to be injected.

The pressure regulator is an integral part of the throttle body. It regulates the pressure by means of a diaphragm-operated relief valve which balances fuel pressure against atmospheric pressure and maintains a constant pressure drop across the injectors. The nominal fuel-line pressure is maintained by the pre-load of a metal spring.

Any amount of fuel not needed to maintain the constant pressure drop is bled off into the return line to the tank.

The two injector valves are electronically actuated on command from the control module. The valve body contains a solenoid whose core piece is a plunger that is pulled upward by the solenoid coil when the latter is energized. Raising the plunger enables the spring to push the ball-check valve off its seat, and fuel begins to flow through the valve.

The quantity of fuel injected depends solely upon the valve-opening duration.

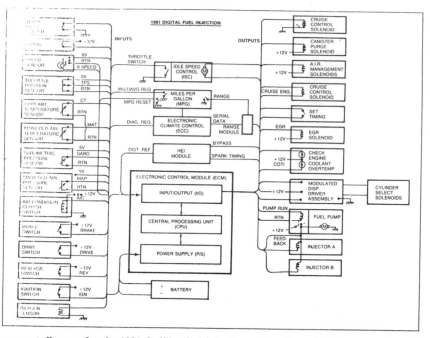

Component diagram for the 1981 Cadillac fuel-injection system shows that the same electronic control unit also handles a lot of extraneous information for control tasks unrelated to the injection system. Sensors that most directly affect the fuel metering are the manifold air pressure sensor, the manifold air temperature sensor, the throttle position sensor, crank sensor, speed sensor, coolant temperature sensor and barometric pressure sensor.

During cranking, both injectors are actuated simultaneously to provide a rich mixture for starting. In normal operation, the injectors are actuated alternately in time with each reference pulse from the ignition distributor.

The intake manifold has a specific design for use with digital fuel injection, and has a built-in exhaust-heat crossover passage for the mixture-enrichment heat-riser valve.

The throttle-position sensor is a variable resistor wire mounted on the throttle body and connected to the throttle valve shaft. The sensor continuously measures the angle of the throttle blades and sends an electrical signal to the control module, which needs the data for ordering mixture enrichment during acceleration and to operate the idle-control system.

The manifold absolute pressure sensor is similar to the unit used with the analog control module, varying the pulse width of its electrical signal in accordance with changes in pressure.

The manifold air temperature sensor is a thermistor whose resistance changes as a function of temperature. The resistance goes up when the temperature is low, and is diminished when the air temperature goes higher.

The coolant temperature is also monitored by a thermistor, inserted just below the thermostat.

The electronic control module is a black box that few of us have any reason to look inside of. What we should know about it can be summed up in a description of its main functions.

The data sensors supply analog signals to the control module which contains a battery of input/output devices that convert the input data into digital signals for the central processing unit, which can handle digital information only.

Control unit for the digital injection system, despite handling a host of extraneous functions, is by far the smallest of the three. At left, the analog control unit developed by Bendix in 1972-73. At right, an interim digital-control design from 1976.

It is the central processing unit that performs the mathematical calculations and logic functions required to produce the correct air-fuel mixture at all times. It also calculates spark timing and idle speed, and commands the operation of emission-control systems, power-train diagnostics and cruise-control.

The control module contains a programmed memory which enables the central processing unit to carry out all these tasks. Electronics buffs will be delighted to know that three differents types of memory are at work in the system.

The first is the read-only memory (ROM), and the program in the ROM is unchangeable. The program is retained if battery voltage is lost or interrupted.

The second is random-access memory (RAM), which works as a scratch pad. Data can be programmed in or taken out freely. The RAM is used for storing rapidly changing data, such as signals from the engine sensors, diagnostic codes and the results of calculations. If a break in battery voltage occurs, all the data in the RAM is lost.

The third is programmable read-only memory (PROM) which contains engine calibration data relative to the engine, transmission, final drive and body type of the car in which it's installed. It is easily programmed, and the data storage is permanent and unaffected if the battery is disconnected.

The present Cadillac system combines the proven analog portions of its predecessor with a digital dioxide closed-loop circuit.

Along with exhaust-gas recirculation and auxiliary air injection, the closed-loop feedback is a function eminently suited for routing through the electronic control unit, which can instigate mixture variations based on oxygen concentrations in the exhaust gas.

Bendix began working on this in 1970, and secured full patent coverage in 1974. The first auto manufacturer to use it was Volvo, using Bosch-made parts, on its 1977-model 2.1-liter four-cylinder 240-series cars with Bosch D-Jetronic fuel injection—for sale in California.

The system includes closed-loop feedback with the usual zirconia sensor placed in the exhaust pipe, upstream of the three-way catalytic converter.

It generates a very weak voltage which varies according to the dioxide content in the exhaust gas. High oxygen levels produce a lower voltage, causing the control module to keep the injectors open longer, thereby giving a richer mixture. Rising voltage output from the sensor indicates low oxygen levels, or a rich-enough mixture. This sensor needs some time to warm up before it can send accurate signals, which limits its usefulness, particularly in the critical warmup period after a cold-soak.

Also, it is strictly an emission-control device, for it is not able to send any warning of overrich mixture, which would be a help toward improving the car's fuel economy.

As early as 1974, Bendix Vice President William L. Miron stated: "Diagnostics are regarded as mandatory because the service industry as we know it today cannot be expected to maintain and repair the sophisticated circuitry, sensors and other components which comprise an advanced electronic control system. But prototypes of these diagnostic systems already are in existence. For example, in the case of electronic fuel injection, a major companion program at Bendix has been the development of appropriate diagnostic tools in recognition of the fact that a new product is useless to the industry if it cannot be easily maintained."

The part-integrated digital injection system of the 1981 Cadillac is the first to offer a true built-in diagnostic routine. It has four programmed types of tests that can be made whenever the situation calls for it. The first type is the engine malfunction tests, detecting system failures or abnormalities. When a malfunction occurs, a warning light goes on and a 'check engine' readout appears on the instrument panel.

The control module memory contains a number of trouble codes for each type of malfunction. The other three types are switch tests, data-display tests and output-cycling tests.

The switch tests check the operation of the switches that provide input to the control module. Data displays are programmed for comparison with an engine that is operating normally, and the output-cycling tests give a check on solenoids and lamps.

If a data sensor fails, the control module inserts a 'fail-soft' corrective value in its calculations and continues to operate the engine. If the defect clears up by itself, the warning light and 'check engine' instruction are turned off, but the trouble-code remains active in a condition known as 'intermittent failure.'

No one at Bendix will pretend that Cadillac's present digital control system is the ultimate form of fuel injection. It is to be regarded as an interim system, but it has everlasting importance in representing a major advance in the state of the art.

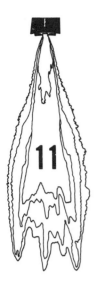

Bosch L-Jetronic

The system known today as L-Jetronic is the direct descendant of the first type of electronic fuel-injection systems developed by Bosch in the 1966-68 period, under Bendix patents.

By Bendix's definition, electronic fuel management is the use of electronics to sense an engine's fuel requirements as a function of measured engine conditions; to compute the amount of fuel needed to satisfy those requirements; and to control the fuel flow accurately in proportion to the air intake for the most desirable combustion level.

Quite early on, Bosch understood that its traditional fuel-injection systems could not compete against this new technology— in fuel-metering accuracy, in emission-control or in providing driveability. Additional pressure came from Volkswagen, fearing that the new antipollution laws in America could result in barring its carburetor-equipped cars from the company's main export market. Volkswagen was well aware of the Bendix developments, to which Bosch had bought the rights in 1965, and asked Bosch to come up with a solution. The experimental departments of both companies adapted the Bendix electronic fuel-injection principles to the air-cooled flat-four VW 1600 (Type 3) engine in record time; thus creating the system now referred to as the D-Jetronic.

Their methods are worth a moment's attention. Fuel requirements for steady-state operating conditions were established by exhaust-gas analysis. Out of these results came the criteria for fuel metering by means of an electronic 'brain.' Auxiliary systems, such as mixture enrichment for cold-starting, warmup and full-load acceleration were added on, along with fuel shutoff on deceleration.

The actual hardware corresponded in broad strokes to that of the analog-computer, timed port-injection system Bendix was then concocting in America.

The VW/Bosch system relied on manifold vacuum sensors for air metering and load measurement, pressurized fuel running in rails and metered by solenoid valves at each injector, excess fuel flowing back to the tank via a return line.

Bosch got a big advantage over all its European rivals by obtaining access to Bendix technology. Two other electronic injection systems from the same period fell by the way-side, partly for lack of know-how, partly for lack of resources.

A French inventor, Louis Monpetit, took out a series of patents for a system he called Sopromi in the years 1967 to 1973. Monpetit used an electronic control unit for actuating solenoids mounted on port-type injectors. The control unit received signals from a mechanical rpm sensor mounted on the camshaft, a coolant temperature sensor mounted on the water-pump housing, a manifold vacuum sensor and a barometer to

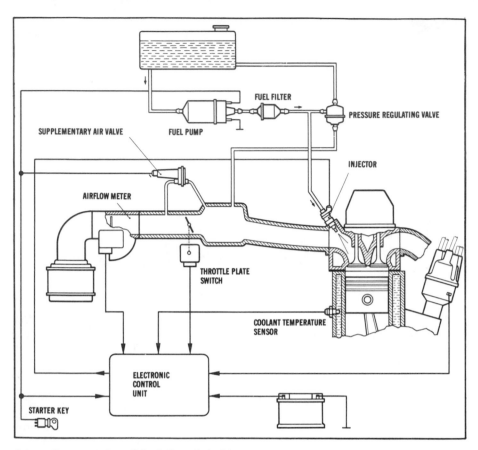

Schematic presentation of the L-Jetronic ignition system.

indicate atmospheric pressure. But the French auto industry did not take the Monpetit Sopromi system quite seriously.

In England, Associated Engineering Ltd. presented its Brico electronic fuel-injection system in 1966. It worked by timed fuel delivery into the ports, with solenoid-operated valves whose opening duration—dictated by electric pulses from the control unit—determined the amount of fuel injected. The electronic control unit was made of two sections: a pulse generator on one side and a combined computer and discriminator on the other, using transistors and printed circuits.

An engine-driven fuel-feed pump delivered fuel to a float chamber which ensured a constant supply for the electrically driven high-pressure pump that kept the fuel circulating in a ring-main under 25-psi pressure.

The injector nozzles were connected in parallel across this ring-main, injecting a conical-pattern spray into the inlet ports. Fuel metering was based on measurements of manifold absolute pressure and combustion-air temperature.

By 1969, however, Bosch had captured the lion's share of the market with its Jetronic (later D-Jetronic). Beginning in 1973, it was phased out in favor of the L-Jetronic, which took advantage of the tremendous progress made in semiconductor technology over the preceding ten to twelve years.

As an example of what this could mean in practical terms, the L-Jetronic control unit has only about eighty components, compared with over three hundred for D-Jetronic.

Main components of the L-Jetronic system. Four injectors (front, left), throttle valve switch, fuel filter and fuel pump (right). The airflow meter and control unit are in the background. In between, from left, are the supplementary air valve, electric starting valve, temperature sensor and fuel pressure regulating valve.

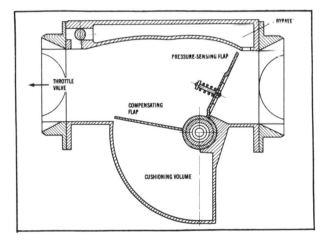

Airflow meter for the L Jetronic system.

It was made possible by such sweeping changes as the replacement of three extensive component groups into simple integrated circuits.

Apart from the three integrated circuits, it contains only a few semiconductor elements plus a number of condensers and equalizing resistors. The unit is connected to the main wiring harness by a multipole plug.

This circuit technology offers extreme precision and a high degree of reliability. It also offers wide flexibility in terms of input and output quantities.

Input signals imparting information about the temperature of the incoming air can be introduced, and temperature compensation can be included by means of a simple resistance network. The system has provisions for additional output signals, concerning

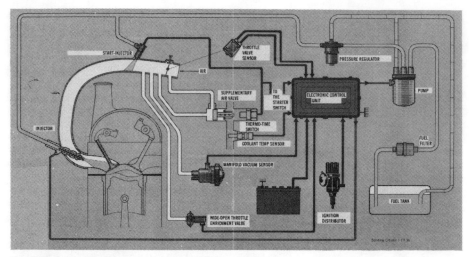

Bosch D-Jetronic system was based on Bendix patents.

Bosch injector contains a solenoid valve, the needle serving as an armature and the magnetic windings being carried in the valve body. Pulling the needle off its seat to about 0.006-inch clearance opens the flow to the nozzle. Opening and closing times are about one millisecond. a-Nozzle needle, b-Armature, c-Coil spring, d-Magnetic windings, e-Fuel entry.

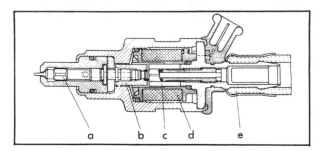

Bosch supplementary air valve works with a bimetallic spring that is electrically heated. It is mounted on the engine at some location where its operating temperature prevails, and opens to admit supplementary air whenever required. a-Bimetallic spring, b-Rotary valve, c-Pivot axis for valve.

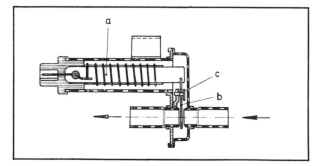

the control of an exhaust-gas recirculation valve, for instance. Also, Lambda-Sond control (oxygen-content feedback from the exhaust system) can be integrated without major complication or expense.

The methods of fuel metering and injection in the L-Jetronic system correspond to those developed for the earlier D-Jetronic. The pressure regulator, coolant temperature sensor and throttle switch are also similar.

In common with D-Jetronic, L-Jetronic is a system which provides intermittent injection into the intake ports at low pressure. Unlike D-Jetronic, however, the L-Jetronic system relies on mass airflow metering as the main input variable to determine the amount of fuel to be injected.

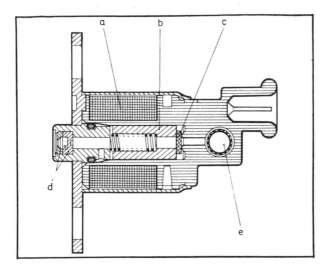

Cold-starting valve contains a simple solenoid valve, spring-loaded in closed position. No rapid-action demands are placed on this valve which is energized via the starter motor if the temperatures are low enough to let the control unit open the circuit to the valve. a-Magnetic windings, b-Armature, c-Seal, d-Nozzle, e-Fuel line.

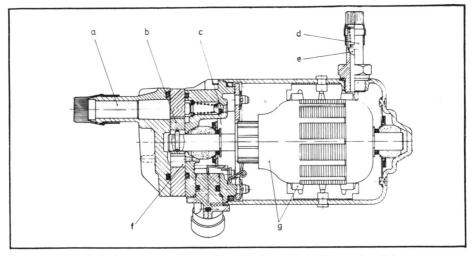

Fuel pump for the D-Jetronic is a roller-cell pump with a cylindrical space in which an eccentrically mounted disc assures the pumping action and builds pressure in the fuel line. Its rotation produces periodic volume variations at the inlet and discharge ports, while the rollers are forced outward by centrifugal force and form a running seal. The rotor is driven by an electric motor. a-Fuel entry, b-Eccentric disc, c-Pressure limiter valve, d-Discharge port, e-No-return valve, f-Metal roller, g-Permanent magnet motor.

In the D-Jetronic system, air metering was accomplished by a manifold pressure sensor, on the principle that while ambient atmospheric pressure prevails in front of the throttle valve, a vacuum is created behind the throttle. This manifold vacuum varies according to throttle position, and can be used as a parameter to determine the load on the engine as well as the volume of combustion air being admitted into the cylinders.

Bosch subsequently established that manifold pressure is only an approximate measure of the amount of air an engine is consuming.

Airflow metering offers vital advantages for mixture control. It compensates automatically for differences in cylinder filling, which may have arisen due to manu-

The pressure sensor contains an inductive data transmitter connected to an electronic time switch in the control unit. Two evacuated aneroids serve to move the plunger in the magnetic circuit of the transformer and thereby change its inductance. With a closed throttle, the manifold absolute pressure is low, and the aneroids (1) are expanded, moving the plunger (2) out of the magnetic circuit. The inductance is low, giving a short pulse (4).

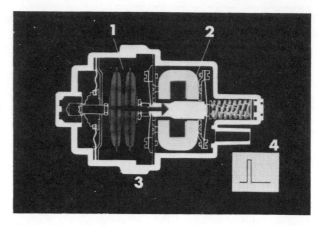

facturing tolerances, wear, combustion chamber deposits or alterations in the valve adjustment on an engine.

It compensates automatically for differences in cylinder filling due to variations in engine speed, and it compensates automatically for differences in exhaust system back pressure (as caused by catalytic converters).

The airflow meter was developed as an independent element, an interchangeable assembly, that is mounted upstream of the throttle valve. It consists of a box on the air duct, with a pivoted flap standing up in the air passage, yielding to the force of the air pressure. The pivot shaft also carries a potentiometer, which converts the flap angle into an electric direct-current voltage signal.

The relationship between mass airflow and the angle of the flap was chosen so that the percentage error is kept constant over the full operating range of the engine. The flap is hinged so that the opening area is increased when airflow increases. It is designed to counteract vortice formation and induced pressure losses throughout its travel.

To avoid overreactions (flagging) in response to sudden changes in the airflow, the flap is spring-loaded against the force of the air pressure. The coil spring has a constant rate, giving a constant balancing force and constant pressure loss.

The flap is also provided with a damper in the form of a compensating flap, connected by a narrow aperture to a damping chamber. It is mounted on the same pivot shaft as the air metering flap. It has the effect of minimizing the effects of fluctuations in manifold pressure on the flap angle. To avoid damage to the airflow meter in case of backfiring in the manifold, the flap has a built-in spring-loaded back-pressure valve.

The relationship between the mass airflow and the flap angle is logarithmic, which ensures a nonlinear characteristic in the potentiometer. The logarithmic relationship obtained this way has the advantage of making the airflow meter most sensitive when airflow values are lowest and the highest precision is needed.

The potentiometer offers the additional advantage that the relationship between the flap angle and the voltage can be expressed in a nonlinear pattern, reflecting the actual fuel requirement by choosing the correct geometry for the track.

The track of the potentiometer is arc-shaped with a ratchet-like contour. Its geometry was determined so as to give an inverse relationship between airflow and voltage. Low mass airflow gives high voltage, and high mass airflow gives low voltage.

Because of the exact signal processing in the control unit, the fuel quantity discharged is inversely proportional to the potentiometer voltage. That necessitated the use of a potentiometer with an exponential curve, which caused a steeper curvature of the voltage and made it necessary to split the track into segments with a parallel, low-resistance series at the junctions between the segments to specify the voltage for the end of each segment. That explains the ratchet-like contour.

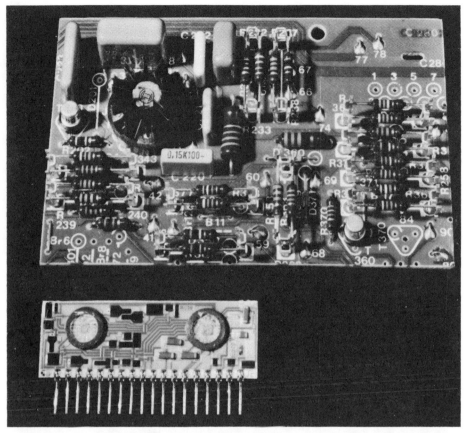

Electronic control unit for the Bosch L-Jetronic system. Data from the airflow meter go into a multivibrator which also receives data on engine speed from an impulse former built into the control unit. The combined data are fed to an amplifier stage, which also receives data on throttle valve angle and temperature. Its output goes to the final stage which arranges the output and commands the injectors.

The series consists of a series connection of cermet (half ceramic, half metal) resistors of high temperature stability. The voltage relationships at the series divider are so far out of proportion to any temperature or aging effects that might occur at the plastic surface, that the series can be optimized strictly on the basis of maximum durability and minimum noise.

The relatively flat cermet resistors can be trimmed to their final values by a computer-controlled laser beam.

These values vary from engine to engine and are stored in the computer, and if the car manufacturer wants to change them, they can easily be programmed from the computer keyboard.

This type of circuit opened the way for combining the notched track with the series divider into one thick-film circuit. All resistance and conductor circuits are combined on a common substrate so that difficult or unreliable soldering points do not exist. This degree of dependability has been obtained only by using the most advanced techniques.

Bosch engineers took extra pains, also, to test the potentiometer, expecting problems with both noise and wear. However, by using a conducting type of plastic with a

The little black box at the bottom shows the central processing unit in actual size. The enlargement above shows its circuitry on a scale that reveals the terrible complexity of electronic data processing.

low friction coefficient in combination with silver-palladium contacts, wear was eliminated as a problem, even at high temperatures.

The voltage from the potentiometer is received in a multivibrator which also receives signals from the impulse-former and ignition contact breaker. The signal from the throttle valve and those from the starting switch and temperature sensor go into the multiplier stage, where they are compared with the input from the multivibrator. The multiplier's output goes into the terminal stage, which directs the injector valves.

Experimental airflow meter of the hot-wire type. It was developed by Bosch and promises to eliminate the mechanical complications of production-type airflow meters.

The voltage from the airflow meter is a measure of air quantity per unit of time, and this voltage determines the frequency for the multivibrator, which is triggered twice per camshaft revolution.

The multivibrator contains a condenser which, during the interval between two firing impulses, is charged at a constant rate so that at the end of the period, the condenser voltage is inversely proportional to the rpm.

The system can handle mass airflow variations over a scale of 40:1 and rpm variations over a scale of 10:1. But the amount of fuel injected varies only over a scale of 4:1.

The airflow meter in L-Jetronic is similar to the carburetor in that it provides a highly accurate measure of the actual mass of air entering a particular engine under any combination of speed and load. Like a carburetor, it automatically responds to the little differences in combustion chamber volume, port shape and other factors which affect volumetric efficiency.

That is good for both fuel economy and emission control. But it has the same weakness that is also inherent in the carburetor: Compensation for changes in air density is not complete, and that entails a tendency to run overrich at high altitude or when the under-hood temperature is high. Theoretical mixture strength is proportional to the square root of air density, and all systems relying on airflow metering are subject to altitude effects. Because the air is less dense at higher altitudes, the mixture tends to get richer.

However, for altitudes up to 7,000 feet this has no practical significance. But there is a problem in that U.S. emission-control standards specify high-altitude tests. Therefore, the system includes an altitude meter, in the form of a barometer converted to potentiometer. Its signals are fed to the control unit, which then adjusts its data from the airflow meter to compensate for the altitude change.

On the fuel-supply side, L-Jetronic has a number of clever features, too. An electric fuel pump delivers the gasoline via a filter to the pressure regulator at a slight overpressure. The fuel pump is of the roller-cell type, driven by a permanent magnet and motor.

The pump itself consists of a cylindrical cavity in which is placed an eccentrically revolving plate. The circumference of the plate has indentations for the metal rollers that serve as seals (due to centrifugal force). That creates a pumping action, since the volume alternates from maximum to minimum as the eccentric plate revolves.

The pressure regulator is a metal casing, divided into two chambers by a diaphragm, which is spring-loaded against the direction of fuel entry by a coil spring. Pretension on the coil spring determines the basic fuel-line pressure. The spring chamber has a vacuum line to the manifold. This serves to make the fuel-line pressure dependent on manifold pressure, so that the pressure loss is uniform for all injector valves. When

The L-Jetronic system is perfectly adaptable for inclusion of the Lambda-Sond, which measures oxygen content in the exhaust gas. This allows closed-loop engine control, continuously adjusting the air/fuel ratio in accordance with the composition of the combustion products.

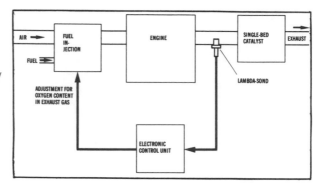

the pressure exceeds the prescribed value, the diaphragm opens an axially adjoining return line, which is not pressurized.

Because the pump constantly delivers more fuel than the engine ever needs, there is a steady overflow being returned to the tank from the pressure regulator.

Fuel metering is accomplished by electromagnetically operated injectors. Branch lines run from the pressure regulator to the individual injectors. At constant fuel pressure, the amount of fuel injected is proportional to the injector-opening duration. This duration is calculated with a high degree of precision and optimized for every engine condition by the electronic control unit.

The control unit receives a continuous flow of signals from sensors that measure charge air temperature, coolant temperature, engine speed and load.

The injection pressure is controlled by the overflow pressure regulator which is balanced on the constant differential between manifold pressure and fuel-line pressure, generally about 35 psi.

The spring chamber in the pressure regulator is connected with the intake manifold. By this method, it is ascertained that the amount of fuel injected depends only on the duration of injector-valve opening.

The injector consists of a body with a jet needle and electromagnet. The electromagnet is shaped like an armature and placed midway in the body. When the magnet coils receive no current, the jet needle is held against its seat by a coil spring. When the current arrives, the magnet is activated and lifts the jet needle about 0.15 mm from its seat, letting fuel run out through a calibrated annular void. The tip of the jet needle is shaped for spraying the fuel in a pattern for best atomization. Opening and closing times for the needle are about one millisecond.

In order to miniaturize the electronic switching, the injector valves were arranged in parallel, so that they open and close, all at the same time, and not in sequence (such as firing order). As a result, there are variations from port to port in the residence time of the fuel injected in each port.

To counteract this, and try for fully uniform fuel distribution, the injectors are set to deliver the metered amount in two portions: half the amount twice per revolution of the camshaft.

This solution avoids any fixed relationship between the cam angle and the moment of injection, and permits triggering of the injection by the contact breaker, so that for every second firing (in a four-cylinder engine), there is one injection impulse.

Some remarks about the injector installation may be important. Fuel deposits on the manifold walls must be avoided in the interest of effective emission control. Therefore it is not advisable to spray against the walls, but to keep the spray angle within twenty-five or thirty degrees, aimed right at the back of the intake valve head.

The injector valves must have thermal insulation against the manifold to avoid hot-starting problems as a consequence of fuel evaporation in the port areas.

The electronic control unit calculates the fuel quantity on the basis of all input data. The order is given to the fuel-injector valves in the form of electrical pulses that actuate them.

These pulses determine the fuel flow per stroke, while the airflow meter indicates mass airflow per unit of time only. The two inputs are coordinated by a division according to rpm. This operation needs a little explanation: Trigger impulses are taken from the contact breaker (or a corresponding terminal with breakerless ignition) and led to the multivibrator via an impulse-former and frequency-splitter stage.

The multivibrator frequency is proportional to the interval between two firing impulses, or in other words, inversely proportional to the engine speed. Consequently, the mass airflow signal must be divided into the rpm, and converted into a number giving air quantity per stroke.

On their way through the control unit, the pulses from the contact breaker are first transformed from peaks into rectangular pulses. These pulses have sharp edges that switch the charging and discharging current of a capacitor in alternation. The charging current is constant, while the discharging current is determined by the angle of the lever on the airflow meter.

It is the discharge period that emits the next switching pulse. Its duration is dependent only upon the interval between two ignition pulses and the ratio of charging to discharging current. This solution assures that engine speed is measured in a mathematically correct mode, and precludes interference caused by temperature fluctuations in the capacitor and voltage variations in the current.

In a second stage, a similar charging and discharging process takes place with these switching pulses. Charging takes place during the pulse, and the charging current is determined by the engine's operating conditions. A resistance wire which senses engine temperature and a throttle switch that signals the load on the engine are connected so as to influence the charging current.

A pulse is also taken at the time of discharge. The discharge pulse immediately follows the charging pulse, reflecting vital data such as load and temperature. The entire pulse goes through an amplification stage, and then its signal goes to the fuel-injector valves.

The majority of all active components are in operation on an on/off basis, which precludes sensitivity to ambient temperature and effects of aging.

The operating mode is particularly well suited to the primary function of the control unit: to conduct the switching operations triggered by ignition pulses through several converter stages directly to the fuel delivery valves, determining their opening duration and thereby controlling the amount of fuel to be injected.

L-Jetronic automatically copes with transitions, such as when the throttle valve is suddenly opened wide and the engine is faced with the risk of fuel starvation. Because of the immediate reaction to any change in mass airflow, the system needs no additional devices to assure adequate fuel supply under those conditions.

Mixture enrichment for cold starts is provided by a magnetic valve located centrally in the manifold.

The starting valve is actuated when the engine temperature is below -15°C, coming into action simultaneously with the engagement of the starter motor. A thermo-time switch turns the starting valve off when fuel enrichment is no longer needed. Because of the starting valve, quick and sure starts can be made in ambient temperatures down to -30°C.

The cold-starting valve is not made for accurate timing but for creating the finest possible fuel spray. The nozzle is of the rotating type and has two tangential bores which promote rotation (by fuel flow). The nozzle gives a fine spray of conical shape, with a spread of no more than forty-five degrees. It has a solenoid valve that is spring-loaded to the closed position. Actuation of the solenoid frees the valve seat and fuel flow begins. The starting valve is disconnected after about thirty seconds. After that, the control unit provides sufficient enrichment.

Impulse diagram for the L-Jetronic system.

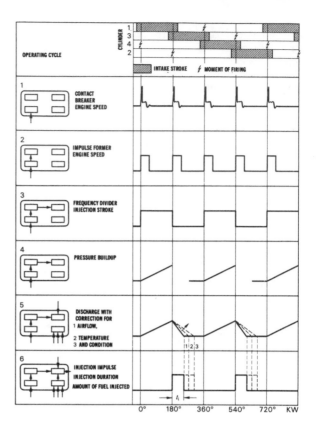

Close control of fuel enrichment during warmup is assured by a simple temperature sensor, mounted at a suitable position in the water jacketing or in the cylinder head, and connected to the electronic control unit.

With very cold charge air, a leaning-out of the mixture occurs due to the physical conditions in the airflow meter. This can disturb an engine, unless the maker has provided preheating of the intake air. For such cases, a temperature sensor can be placed in the intake duct, to deliver a corrective signal to the control unit.

On cold starts and during warmup the engine needs auxiliary air for idling due to the higher friction losses that are at play. It becomes necessary to assure a faster idle. Additional air is supplied via a bypass valve which leads a small amount of air around the throttle valve. Its opening area is controlled by engine temperature and the temperature of an electrical heater coil with a bimetallic strip. This additional electrical heating is intended to speed the warmup and shorten the time when enriched mixture is supplied.

On the overrun (when coasting or using the engine as a brake) the engine places specific demands on the air/fuel ratio, which are different from the conditions when the engine is pulling.

L-Jetronic can be set to cut fuel delivery entirely above any chosen rpm, and to give a leaner mixture at other speeds. The only condition is that the control unit must be fed information about mass airflow and rpm in the proper context.

To assure correct engine operation on the overrun at very low car speeds, L-Jetronic opens the bypass valve to admit additional air past the throttle valve and prevent an excessive pressure drop in the manifold.

In many cases, it is preferable to adjust full-load and idle settings separately from the broad spectrum of operating conditions. This is possible with L-Jetronic, which has a throttle switch that signals the control unit whenever idle or full-load conditions exist, so that the air/fuel ratio can be optimized for those exact conditions.

For idle adjustment, the system even includes an adjustable bypass to the airflow meter.

For safety reasons, L-Jetronic contains an electrical cutoff which prevents fuel delivery when the ignition is turned on but the engine not running. This is achieved by a switch that cuts the current supply to the fuel pump when the flap in the airflow meter is fully closed. The switch is overridden by engagement of the starter motor.

In engines whose exhaust is treated in a catalytic converter, it is not possible to limit peak rpm by inserting a centrifugal governor in the ignition distributor, because failure to fire would cause the unburned mixture to overheat the catalyst.

An upper rpm limit is assured in L-Jetronic by simply giving orders to the control unit to restrict fuel injections to a certain maximum frequency. This gives an effective speed governor without risk for the catalytic converter.

All data regarding the engine functions are collected in the control unit, and they can be made to serve additional purposes. The control unit can be wired to give an audio or visual warning signal to the driver about overspeed, wasteful fuel flow rates and over-heating.

The small dimensions and light weight of the L-Jetronic components make it easy to install in a car. Attention must be paid to provide the airflow meter with rectifier stages on both sides so as to maintain closely defined flow conditions inside it. Otherwise the engine manufacturer has full freedom to put the elements where most convenient, and to optimize the air intake duct for ram effect.

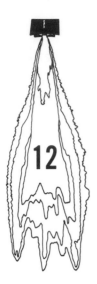

12

Bosch K-Jetronic

The name is perhaps misleading, because the Jetronic trade name was identified with electronic systems before the introduction of K-Jetronic in 1973. K-Jetronic is not electronically controlled, but is a mechanical system. In fact, it is one of the simplest of existing non-electronic injection systems. It differs from other mechanical injection systems on three important points:
1. Fuel is injected continuously.
2. No mechanical drive is used.
3. Fuel metering is based on mass airflow metering.

Looking at all possible means of eliminating the mechanical drive, Bosch engineers reasoned that the best way would be to adopt continuous injection, which meant developing a method for continuous measurement of the mass airflow as well as continuous fuel metering.

Mechanical control systems are not capable of taking into account the air-consumption tolerances of the engine, nor can they allow for the effects of partial exhaust-gas recirculation (as needed in many emission-control systems). Measuring the mass airflow, on the other hand, would—theoretically, at least—enable these factors to influence the fuel metering to the proper extent. The validity of the theory was proved in basic experiments.

Continuous injection was chosen, also, because a mechanical system is not easily allied with electrical triggering of the injection, which could be done, for instance, by wiring the system to the ignition distributor to get an rpm reading.

The Bosch engineers reasoned further that the essential ingredient for proper control of air/fuel ratios over a wide band of operational conditions was to measure the flow of air drawn into the engine.

It can't be measured by weight, but the flow in terms of cubic feet per minute can be accurately measured. For proper engine operation, however, the numbers are not relevant. All that is needed is a device that accurately transmits a signal, on a scale arranged in linear progression as the mass airflow increases. That is basically what a carburetor does, but as we shall see, K-Jetronic is not a type of carburetor.

By measuring the airflow mass, the Bosch engineers chose the control principle that promised to give the closest conformity with the actual goings-on in the engine, so as to offer optimum opportunity for controlling exhaust emissions.

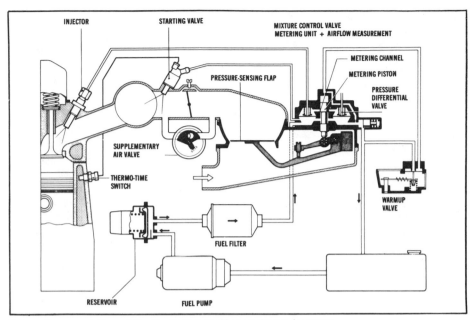

Schematic presentation of the Bosch K-Jetronic non-electronic system.

Mass airflow is proportional to the cross-sectional area open to it, the air density and the velocity. According to the proven formula, M = pAV when A is the annular area around the plate, V is velocity and p is air density.

With exact measurement of the volume of air flowing into the engine, the fuel metering could be automated, as it is done in constant-vacuum or air-valve carburetors. That type of carburetor did indeed serve as a model for the first experimental version, called the KA system (K for continuous, A for driveless).

The air-valve or constant-vacuum carburetor works with an airflow range of about 30:1—a maximum flow rate thirty times greater than the minimum. By comparison, the venturi carburetor has an airflow range of 900:1 which precludes the degree of accuracy that Bosch regarded as necessary for a competitive injection system.

A piston arrangement, patterned on that of an SU carburetor, was tried in the first stage of experimentation. Many other configurations were also tried and evaluated. But the piston type was less consistent and less immune to friction effects than the lever system that was selected for further development.

The production-model airflow meter is built around a venturi containing a circular metal plate, suspended on a counterbalanced lever. The suspended plate is forced back by the inflow of air, lifting and tilting from its pivot. The deflection occurs at a linear rate relative to the flow rate, continuously balancing its angle and opening against the force of the airflow.

This linear relationship between mass airflow and fuel metering is the key to maintaining a constant air/fuel ratio regardless of operating conditions in modern engines having a load range of 4:1 and a speed range of about 8:1. After considerable experimentation, Bosch arrived at a device which faithfully translated this relationship into the movement of a single lever.

By making the air passage conical in shape it was assured that the increase in area is exactly proportional to the change in plate angle.

From this fact it follows that the relationship between mass airflow and the cross-sectional area will be in direct linear proportion (since airflow velocity remains constant).

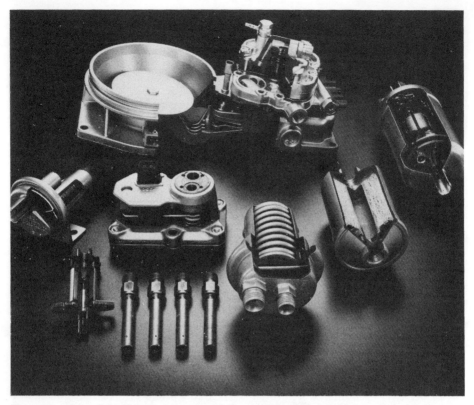

The components of the K-Jetronic system: Airflow meter and throttle body combined in the background. Clockwise from there, fuel pump, fuel filter, reservoir, four injectors, electric starting valve and supplementary air valve. In the middle, the warmup valve.

The pressure on the plate is determined by the airflow velocity and air density according to this formula: $P = \frac{1}{2}p\ V^2$ where P is pressure, p is density and V is velocity.

During test work, Bosch engineers arrived at suitable dimensions for the air meter plate:

60 to 80 mm . . . . . for four-cylinder engines from 40 to 120 hp

80 to 85 mm . . . . . for six-cylinder engines from 130 to 200 hp

110 mm . . . . for eight-cylinder engines up to 300 hp

Increased plate diameter for bigger engines became necessary due to a loss of accuracy when the plate is forced to extreme angles. That in turn imposes a limit on the pressure drop that can be considered tolerable. Tests have shown that with large mass airflow rates and high plate angles, separations occur in the air stream, leading to an unsteady lever.

Opposing the pressure of the airflow against the plate by means of a hydraulic constant-control force will create the sort of situation where any increase in the force of the airflow will deflect the plate through an angle that increases the cross-sectional area of the annular passage around it exactly enough to bring the velocity back to its former level.

A bridge-piece below the plate works an electric switch which breaks the fuel-pump circuit whenever no air enters the duct. Any backfiring through the intake manifold could seriously upset the functioning of the airflow meter. For this reason it has a safety release that allows it to swing over and release the back pressure into the entry duct.

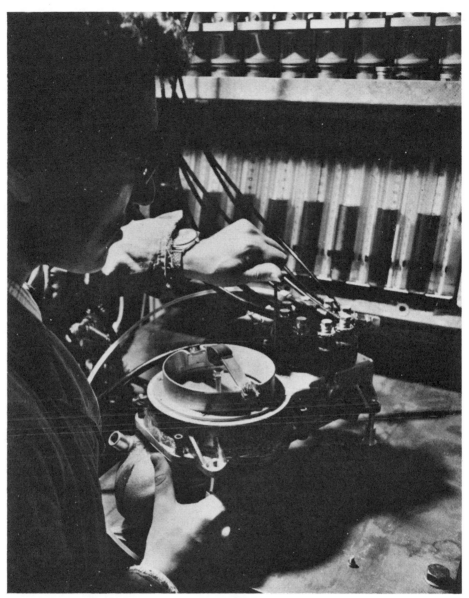

K-Jetronic fuel metering unit being tested at the Bosch factory.

Overcenter travel of the plate is restricted by a rubber cushion, and an adjustable leaf spring sets the plate in its correct neutral position that is necessary for normal starting.

The airflow meter has a body of pressure-cast aluminum. The plate is also made of aluminum, so that the two are fully compatible in terms of thermal expansion or contraction during temperature variations in the charge air.

The venturi design was carefully laid out to make sure that all of the incoming air participates in bringing pressure on the plate. For the same reason, the usual idle adjustment by air bypass with a screw-type throttle valve was discarded in favor of a mechanical

The principle of the disc and roller action in the electric fuel pump is explained in these drawings. Fuel enters at (1) and a quantity is trapped in the cell formed by the eccentric disc and its rollers (3). The volume of this cell is reduced by continued rotor rotation, which raises the fuel pressure and forces it out the discharge port (5) under pressure.

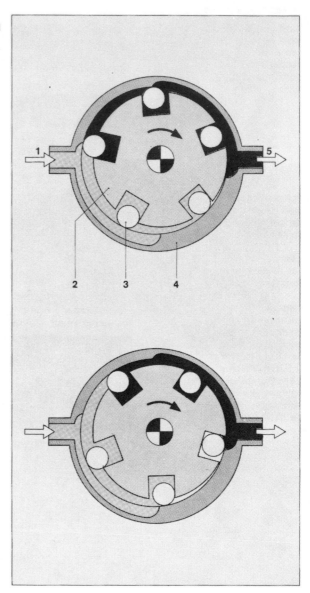

linkage between the air meter and the metering plunger. Under pulsating-flow conditions, sensor plate movement is damped by a fixed-caliper pinhole throttle.

The pivot axis for the plate is a stainless-steel shaft, carried in calibrated teflon-coated bearings. An idle-speed roller is mounted on an intermediate arm that is adjustable relative to the main arm. Idle speed adjustment can thus be carried out with the engine running.

The distance between the bearings on the pivot shaft has been stretched to the maximum, with plenty of axial play. The output roller runs on a stainless needle roller bearing with minimal radial runout. The idle-adjustment screw is self-locking and carries

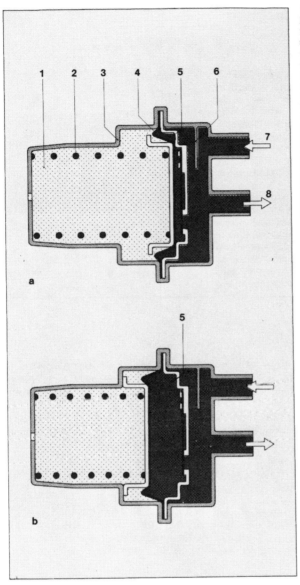

The reservoir has the task of keeping a small amount of fuel ready for facilitating restarts, especially with a hot engine, shortly after switching off. It has a fuel chamber divided by a diaphragm, and a cylindrical chamber containing a coil spring loaded against the diaphragm. Fuel flow puts a counterpressure on the diaphragm when the engine is running. When the engine is turned off, the diaphragm gives in to the coil spring. a-Empty, b-Full, 1-Spring chamber, 2-Spring, 3-Abutment, 4-Diaphragm, 5-Reserve volume, 6-Baffle, 7-Fuel in, 8-Fuel out.

a polished impact surface. The output lever is a generously dimensioned aluminum pressure-casting.

The proportional movement of this lever from the pressure plate can be transferred to a fuel metering plunger by mechanical means, with instant reaction to changes in mass airflow.

The control plunger of the fuel metering unit is under hydraulic pressure from above, working in a sleeve or barrel provided with a number of very narrow, vertical slots of rectangular shape. It applies a hydraulic force in opposition to the airflow, finding its position at the point of equality between airflow and hydraulic force. When the lever raises the metering plunger, more fuel is admitted into the distributor unit.

The pressure regulator valve is built into the metering unit. It keeps fuel pressure at a constant 73.5 psi by routing excess fuel back to the tank. Its action is dependent on balancing the fuel pressure against the spring load behind the piston. a-Engine stopped, b-Engine running, 1-Fuel under pressure, 2-Seal, 3-Return line, 4-Piston, 5-Coil spring.

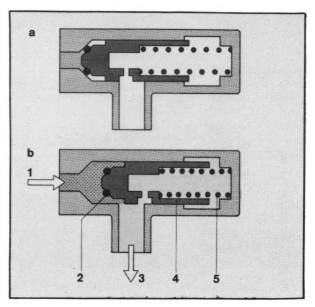

The metering and distributor unit is installed upstream of the throttle valve. It should be positioned horizontally, preferably with flexible mounting, so as to avoid acceleration forces exceeding 5 g. Injectors should be installed as close to the intake valves as possible, according to Bosch experts.

It was a self-imposed condition for the development of the K-Jetronic that each intake port should have individual shutoff control, and operate at a higher pressure than any pressure head that could normally occur in the manifold or port areas. This condition plays on the requirements for reliable hot-starting, for instance.

The reason for wanting the injector so close to the intake valve is that this arrangement permits satisfactory operation with the leanest mixture, and therefore the lowest fuel consumption.

The electrical system is arranged so that the fuel pump begins to function when the starter motor is turned on. When the engine fires, and the starter is disengaged, an electric switch on the air metering lever permits the fuel pump to continue its operation. At the same time the heating of the resistance wires for the warmup valve and auxiliary air valve is initiated.

The start-enrichment valve which is connected to the thermo-time switch in the water jacket, is actuated through the starter motor operation.

K-Jetronic includes two fuel circuits: primary and secondary. The primary circuit is the supply circuit. It begins with a submerged roller-type electric pump which feeds fuel into an accumulator.

Why an accumulator? Bosch engineers had good reason to put it there. It delays pressure buildup during the starting sequence and after the engine has been switched off. It also helps in damping the noise from the fuel pump. From the accumulator, the fuel flows through a filter into the combined metering and distributor unit. Surplus fuel is routed back to the tank via a constant-pressure regulator.

The accumulator is a tank with two chambers, separated by a single-leaf diaphragm. The diaphragm separates the spring chamber from the fuel chamber. Fuel chamber displacement is about 20 cc and the spring is arranged to pressurize the fuel contained there during periods of standstill to a level of 37 to 51.5 psi.

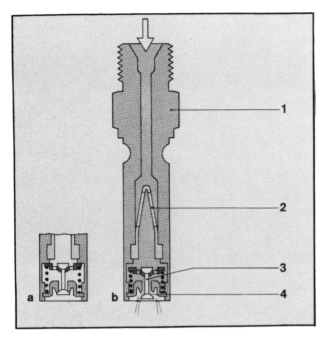

The injector nozzle is installed in a holder offering a high degree of temperature insulation to prevent vapor lock when a hot engine is switched off. The nozzle is shaped to deliver the fuel in a conical spray formation. a-Closed, b-Open (injecting), 1-Holder, 2-Filter, 3-Needle, 4-Valve seat.

A spring-loaded valve with a throttle makes it possible to obtain rapid filling of the fuel chamber without needless loss of the effective content in it when the engine is switched off.

In the metering and distributor unit, both upper and lower body halves are made from cast iron. The diaphragm is made of steel, which has a similar rate of thermal expansion, so as to assure freedom from variations due to temperature fluctuations.

The distributor body contains a set of pressure-regulating valves—one for each injector. This valve consists of two chambers, separated by a steel diaphragm. This diaphragm separates the primary and secondary circuits.

The secondary circuit is the distribution circuit, which starts with the distributor unit and ends with the injectors in the ports. Fuel flows through a ring-channel in the distributor unit with ports to the lower chambers of the pressure-regulating valves.

The sleeve and pressure-regulator valves are made of stainless steel. The plunger has undergone electro-erosive treatment. Inside this unit, there are no adjustment possibilities. Everything must meet specified tolerances in production.

Because of the difference in forces acting on the air meter in four-, six- and eight-cylinder engines even with identical system pressure, Bosch engineers found it necessary to adopt valve pistons of different size:

10 mm .... for four-cylinder engines
13 mm .... for six-cylinder engines
17 mm ....for eight-cylinder engines

By using large-size cross-passages between the individual chambers, the valves operate with exact parallel switching without pressure loss. The entry chambers, following the plunger sleeve, must have sharply defined separations, so as to avoid mutual spill-over. That has been achieved with radial-sealing O-rings for each slot.

To protect the pressure-regulating valves, a nylon filter with a mesh width of twenty-five micrometers is installed prior to the distribution slots.

Pressure in the ring-channel is maintained at a constant 69 psi by a regulator included in the primary circuit. Short tubular valve seats are mounted in the upper chambers

The airflow meter in the K-Jetronic system has a flap mounted on a pivoted and counterweighted lever. Lever motion, modified by the mixture control screw, is transmitted to the metering unit by a plunger-type follower. a-Flap closed, b-Flap open, 1-Body, 2-Flap, 3-No-load area, 4-Mixture control screw, 5-Counterweight, 6-Pivot point, 7-Lever, 8-Leaf spring.

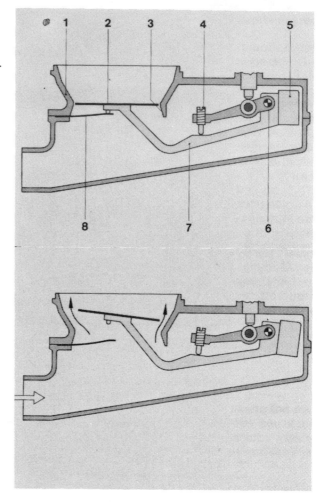

of the pressure-regulating valves of the secondary circuit. Their lower ends bear against the diaphragm, which responds to variations in pressure differentials between the upper and lower chambers. The deflections of the diaphragm dictate the cross-sectional open area of the valve.

By deflections of the diaphragm, the pressure differential between the two chambers is maintained at 1.5 psi, a very low figure which permits extremely small port slots, only 0.1 to 0.2 mm (0.004 to 0.008 inch) in width.

Fuel enters the valve bodies in quantities determined by the metering plunger. The fuel is then admitted through the slots in the metering sleeve to the discharge ports of the pressure-regulating valves.

Pressure in the secondary circuit varies between 7.5 and 51.5 psi. A pressure of 48.5 psi is required to open the valve in the injectors. The injector valves include a miniature nylon fuel filter and a chatter valve that senses low flow rates and starts vibrating to assure adequate atomization of the fuel under such conditions.

The injector nozzle is basically the same as in older injection systems with in-line pumps. It is a forward-opening spray valve with a ball-type seat and spherical suspension

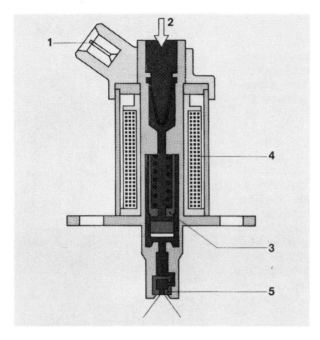

Cold-starting valve in the K-Jetronic system is electrically actuated from terminal (1). Fuel enters (2) and is filtered on its way to the nozzle (5). Magnetic windings (4) pull on the armature (3) to unseat the valve when cold-start enrichment is wanted.

from a spring-loaded base. Each injector is protected by a nylon filter with a mesh width of fifteen micro-millimeters.

By miniaturization of all moving parts, their natural frequency has been raised high enough to permit minute quantities to be injected. At idle, a quantity of 5 cc per minute enters the needle of a reed valve whose natural frequency is 1,500 cycles per second.

Stable and frictionless self-centering of the needle is assured by its short construction and interface with the spring, which is shaped for it. The valves are thermally insulated by O-rings in plastic carriers. Maximum valve temperature does not occur till the engine is switched off, and should not exceed 95°C.

The intake manifold should be laid out in accordance with three requirements:
1. Optimum cylinder filling
2. Uniform mass airflow to each cylinder
3. Adequate length for smooth transitions

To explain point number three: For easy transitions from one set of operating conditions to another, the volume of the manifold section between the intake valves and the throttle valve should be equal to or 1.5 times the engine displacement.

Downstream of the throttle valve, the manifold contains a fuel-spray valve designed to provide mixture enrichment, as necessary, for cold-starting. It is electrically operated by an automatic switch that registers both time and temperature signals.

The cold-start enrichment valve is connected to the primary fuel circuit by open lines. A third circuit has the vital role of providing an input on the metering plunger in accordance with the temperature rise during engine warmup. Pressure in the warmup circuit is controlled by a pinhole throttle connected to the control circuit. This throttle valve is opened and closed by the cold-start enrichment valve.

Fuel flows to the warmup valve through an open line from the cold-start enrichment valve, and a bimetallic spring keeps the warmup valve open during cold-starts.

The bimetallic spring is heated by an electric resistance wire, which makes it return to inactive position within a predetermined time (regardless of coolant temperature) and close the warmup valve.

The warmup valve cuts back on enrichment gradually as the engine builds up to normal operating temperature. With a cold engine (a) the bimetallic spring (5) is electrically heated (6) from the moment the engine is started. It bends down to let the diaphragm (1) yield to control pressure (3) and open a bigger volume opposite the return line (2). At normal temperatures, the spring is raised, and the diaphragm is in neutral position.

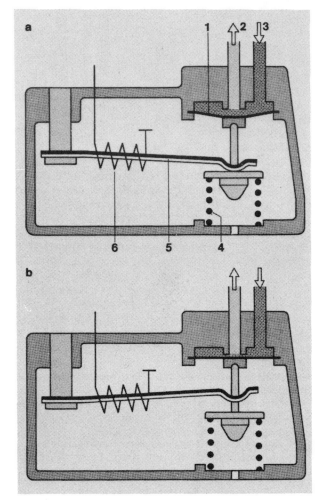

During warmup, the valve relieves hydraulic pressure to the metering plunger, and when normal operating temperature is reached, the valve closes to maintain a constant hydraulic load on the plunger.

Due to its basic principles of operation, K-Jetronic had to operate without the air containment (as used with electronic fuel-injection systems), which meant that different mixture-preparation methods had to be developed.

Air containment assures excellent mixture preparation at idle and in the lower part-load regions, but gives no assistance to mixture preparation with increasing absolute pressures, such as full-load or transient conditions.

Variations in equivalence ratios occur during transitions. For instance, the charge becomes rich after sudden opening of the throttle. That is necessary and corresponds to the action of the acceleration pump in a carburetor. After closing of the throttle but no drop in rpm, the charge becomes leaner. With a sudden drop in rpm on a closed throttle, the charge becomes richer. These changes are mainly due to the conditions existing in the intake manifold between the throttle valve and the intake valves.

The airflow meter accurately registers the change in filling that accompanies a change in load (manifold vacuum). For instance, when the throttle is suddenly opened, the airflow meter must admit the air quantity demanded by the cylinders plus the air quantity needed to raise the manifold pressure to its new level.

The most difficult condition for the system to cope with is a sudden opening of the throttle shortly after a cold start, from low-rpm operation. Under this situation, a large part of the fuel injected will end up as raw fuel, gravitating to the gutters of the manifold. The mixture will be extremely lean. The engine will not accelerate, and flameout may occur. The engine may stall.

In a series of experiments, Bosch has demonstrated that the average deviation from the chosen fuel equivalence ratio can be reduced by two methods. The first is to increase airflow velocity across the airflow meter. The second is to include a damping zone between the airflow meter and the throttle valve. Both methods serve to avoid airflow separation through the airflow meter.

K-Jetronic may have its minor drawbacks, but the system is not running any immediate risk of large-scale displacement by L-Jetronic or newer systems. It is extremely versatile, giving satisfactory results on engines of widely different types, and is fully compatible with turbocharging, for instance. Above all, it has a tremendous cost advantage over all other fuel-injection systems; and for markets with strict antipollution laws, it can compete on equal terms with carburetor systems.

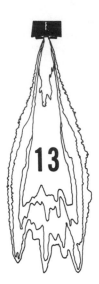

Bosch Motronic

First released by BMW for the 732i model in November 1972, the Motronic system is not a new type of fuel injection, but a conversion of the L-Jetronic to digital operation, with incorporation into the same control unit of the ignition timing. By coordination of the fuel metering and the spark timing, the engine comes under optimal and continuous adjustment with regard to efficient combustion (minimal fuel consumption), the composition of the exhaust gases, the car's performance potential and its driveability.

In previous systems, the ignition was adjusted to give the best-timed spark for sure-firing under all conditions, but this could be accomplished only within the framework of the adjustment angles controlled by the centrifugal and vacuum-operated spark-advance mechanisms on the ignition distributor.

It was not able to take any account of mixture strength. Consequently, the fuel metering system and the ignition system occasionally worked at cross-purposes.

Uniting signals of all engine functions and integrating the control of spark timing and fuel metering in the same microprocessor goes a long way toward keeping the engine optimally adjusted to all possible speed and load combinations.

Despite the switch from analog to digital control, it was possible to retain all the sensors from the standard L-Jetronic. One new sensor was added, the combined rpm-counter and timing-mark sensor. It is an induction-type sensor with a permanent magnet, carrying an annular gear whose teeth modulate the signals from the emitter so as to set up an approximately sinusoidal-wave-induced current.

This current gives an engine-speed reading, which is needed for both fuel metering and spark timing. A pickup on the annular gear (or a separate disc rotating at the same speed) registers with the fixed timing marks in sequence, and produces another induced current with intermittent signals.

The airflow meter itself is the same as in standard L-Jetronic systems. But for the Motronic, instead of a potentiometer whose voltage begins a declining hyperbolic curve as the plate angle increases, the potentiometer voltage undergoes a linear rise in the same situation. This change was made necessary by the adoption of a digital microprocessor.

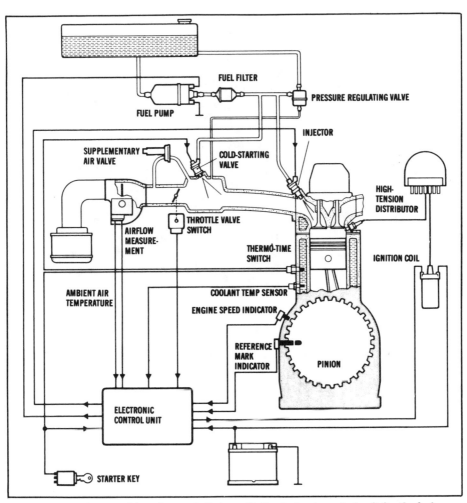

The Motronic system combines the spark timing with the electronic control unit for the fuel-injection system, which is based on the L-Jetronic.

The electronic control unit is composed of a central processing unit that handles the logic and calculations, two memory banks and an input/output cell that handles all communications with the outside world.

One memory is a read-only (ROM) and the other a random-access (RAM). Their tasks are divided in the same general pattern as outlined in the chapter dealing with the Bendix digital injection as adopted by Cadillac.

The input/output cell has the task of preparing the input signals for further processing, so that all output data are delivered in digital form. The central processing unit is not able to understand anything but digital signals.

The conversion produces rectangular signals that carry bits of information in their length, width or frequency. These rectangular signals enter the switching circuit where they are transformed into eight-bit words, ready for feeding to the central processing unit.

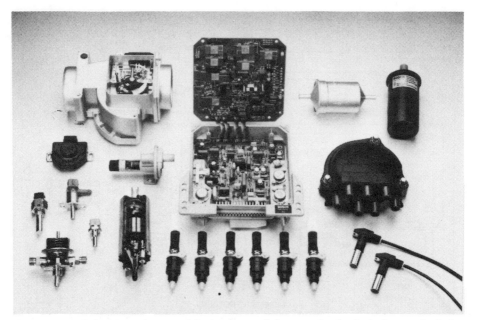

Components of the Motronic system. Grouped around the control unit, clockwise, are the fuel filter, ignition coil, distributor cap, six injectors, fuel pump, temperature sensor and thermo-time switch, pressure regulating valve and cold-starting valve, throttle valve switch, supplementary air valve and the airflow measurement unit.

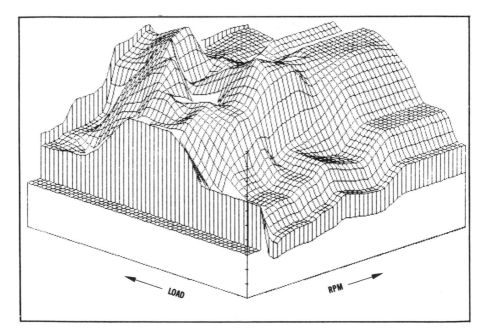

Spark-advance map for the Motronic-equipped engine shows far greater sensitivity to load, up and down the full speed range, and therefore contributes to more efficient combustion (more power for less fuel).

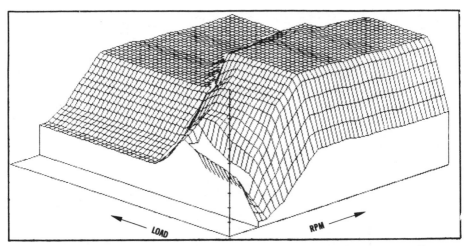

Spark timing with electronic ignition varies in a simple three-dimensional pattern, according to load and rpm.

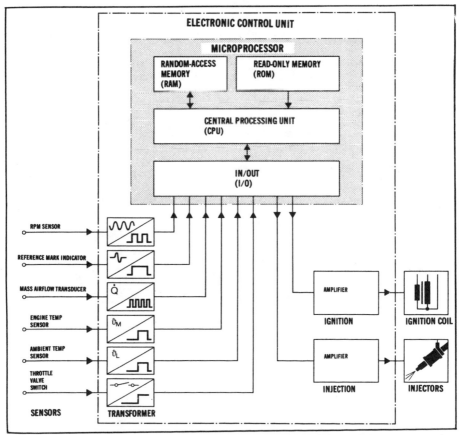

Switching diagram for the Motronic system. Central processing unit is coupled to two memories, RAM and ROM, and an input/output element. Inputs come from (top to bottom): Rpm sensor, reference mark indicator, mass airflow transducer, engine temperature sensor, ambient temperature sensor and the throttle valve switch. Outputs go to injectors and the ignition coil.

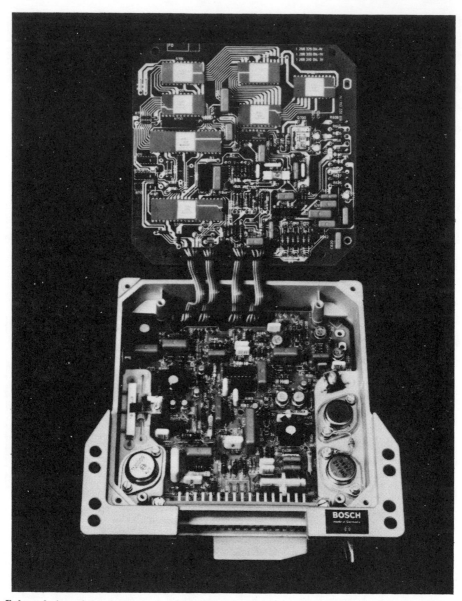

Enlarged view of the electronic control unit for the Motronic system.

The programs stored in the memory banks then come into play, engaging the logic elements whose output is fed back to the switching circuit from where it emerges in the form of physical impulses for fuel metering, injection timing and spark timing. As they emerge, the signals are weak, and pass through an amplifier that enables them to actuate the ignition coil and the injector valves.

The microprocessor, which has complementary metal-oxide-semiconductor (C-MOS) technology, depends for its operation on a current supply of suitably low voltage and a stable frequency giving a constant time reference. The frequency is regulated by a quartz-type oscillator.

As used in the BMW 732i, Motronic is programmed to cut off fuel delivery on a closed throttle. Then, when the engine speed drops down to 1200 rpm, the control unit instructs the injector valves to resume fuel delivery. This is calculated to save between three and five percent gasoline without affecting the car's behavior or the driving tasks at all.

Motronic is totally maintenance-free. Tune-ups are eliminated, and all the car owner has to remember is to change plugs at prescribed intervals. BMW has even developed a system that enables the same microprocessor to calculate plug life on the basis of actual operating conditions plus a time factor, instead of simply counting mileage and giving a reminder in the form of a readout on the instrument panel.

Just like the earlier L-Jetronic, Motronic is perfectly adaptable to various subsystems, such as Lambda-Sond (oxygen-content feedback from the exhaust gas), digital idle stabilization and exhaust-gas recirculation.

Bosch, BMW and ZF are currently collaborating on linking up an automatic transmission to the same microprocessor, so as to obtain electronically controlled shifts. This will bring the overall gearing into the system so as to keep the engine operating at the most efficient rpm versus load combination for all driving conditions.

Bosch Mono-Jetronic

At this writing, Mono-Jetronic is not yet used as standard or optional equipment on a production car. It grew out of a joint research program between Bosch and Volkswagen, the car company being interested in investigating the advantages for the fuel-mixing process of removing the injector nozzles from the port area and relocating them further upstream in the intake manifold.

Coupled with this desire to obtain closer control over fuel atomization so as to prepare a more homogeneous mixture, Volkswagen had the secondary goal of trying out various ideas for reducing the cost of the fuel-injection systems.

The result was to be expected. When the four injector nozzles for a Golf engine were moved far enough back from the ports, they had to be positioned very close together. It became obvious that they could be replaced by one single injector.

From this basic concept, Bosch created a mixing unit with an electromagnetically operated injector valve positioned immediately above the throttle plate.

The mixing-unit body is equipped with a bypass for auxiliary air and also carries a fuel-pressure regulator. The fuel metering is performed electronically with a control unit and sensors similar to those of the L-Jetronic system.

Of course, with single-point injection, there is no need for timing of the fuel delivery. But since the L-Jetronic system works under constant pressure and only the opening duration of the injector valve can produce variations in the amount of fuel delivered, the principle of intermittent injection was adopted also for the Mono-Jetronic.

Early Mono-Jetronic prototypes used the L-Jetronic airflow meter, but later versions have been equipped with a hot-wire sensor that is simpler, smaller, lighter, cheaper and easier to install.

It works on the principle that a flow of air at ambient temperature will have a cooling effect on a heated wire. The temperature drop being proportional to the mass airflow, the resulting reduction in the resistance of the wire can be utilized as a measure of changes in the volume of air admitted to the engine.

Bosch chose platinum for the wire, which was given a diameter of 100 microns and connected to a Wheatstone bridge circuit. The wire runs across a tubular duct in-

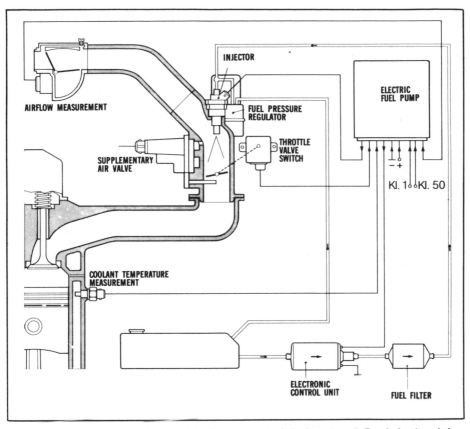

In search of the simplest possible fuel-injection system (and the lowest cost), Bosch developed the Mono-Jetronic. The individual injectors in each intake port have been replaced by a single injector mounted at the entrance to the throttle body. The airflow measurement device is of the same type as in the L-Jetronic system.

stalled centrally in the air-intake duct, closely downstream of the air cleaner. The tube has straight walls, so that it is free of venturi effects.

The airflow meter is mounted ahead of the throttle plate where airflow-velocity readings tend to be highest.

Through the Wheatstone bridge circuit, the wire is heated to a constant temperature of 150°C. Whenever the mass airflow increases and cools the wire, the heating current is automatically increased by a small electronic unit that serves as a continuous monitor of the current and resistance in the wire. Signals indicating the strength of the heating current needed to bring the wire back to its reference temperature are relayed to the main digital control unit.

The hot-wire sensor automatically compensates for changes in altitude, since the wire's temperature will be affected by changes in air density. Bosch estimates that a 3,000-foot difference in altitude can introduce an error of up to five percent in flap-type airflow meters.

Because of its many advantages, the hot-wire sensor is under test in combination also with multipoint injection systems.

Mono-Jetronic is being prepared for commercial production and is expected to become available shortly.

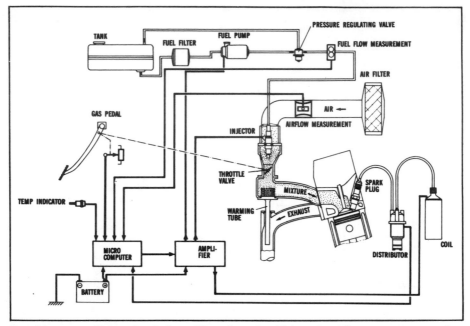

Complete system diagram for the latest Mono-Jetronic with hot-wire airflow measurement and direct preheating of the incoming charge by exhaust heat. The supplementary air valve has been eliminated, and all temporary enrichment demands are satisfied by the single injector.

Lucas Digital Fuel Injection

As early as 1970, Lucas Industries of Birmingham, England, revealed the results of some experimental work involving an analog/digital converter as part of an electronic fuel-management system. It was installed on a Triumph 2500 PI which was also equipped with Lucas electronic antilocking brakes, electronic ignition, electronic cruising-speed control and an electronic monitoring diagnostic system.

Demonstrations for the British motor industry did not lead to any orders, however. It was not really ready for production, and Lucas—with assistance from radio- and telecommunications-giant Ferranti—was already deep into a second-generation digital-control concept. Two important new patents were granted to Lucas researchers Malcolm Williams, Duncan B. Hodgson and Michael M. Bertioli, in 1972.

Initial development work was performed on a Jaguar V-12 engine with a system based on a large-scale integrated circuit capable of handling digital information. Test results were encouraging and, by 1977, Lucas was again making demonstrations to the motor industry, not only in Britain but also in Europe, the USA and Japan.

It is a timed, port-type injection system, whose essential novelty lies in the electronics. In the proposed system, the main printed wiring board carries two large-scale integrated circuits. One circuit handles the fuel metering functions and the basic read-only memory that had been programmed with the fuel schedule. The other circuit receives analog signals from a plurality of transducers and converts the signals to digital form.

The patented elements of the system are concerned with the method of processing input information. Three main parameters are commonly used for fuel metering: engine speed, manifold pressure and throttle position. Transducers provide signals for each parameter, each signal consisting of a three-bit binary word.

In the Lucas system, signals for two of the parameters are selected for processing through a diode core. They are not always the same two parameters—any combination can be used. The three-bit binary words from the chosen two transducers are fed into a pair of decoders. There are eight combinations of each digital signal and, for each input signal, the decoder energizes one of eight input lines to the diode core. Output lines from the decoder control a switching device in the diode core.

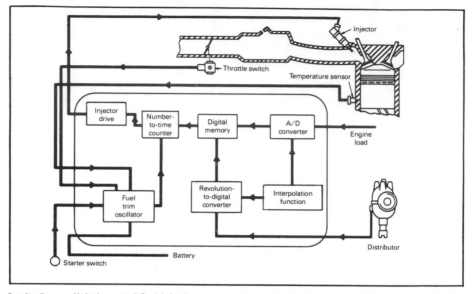

In the Lucas digital-control fuel-injection system, the control unit gets a reading on engine load from a manifold vacuum sensor. Speed readings are taken from the ignition distributor.

Schematic drawing of the input/output signals of the electronic control unit.

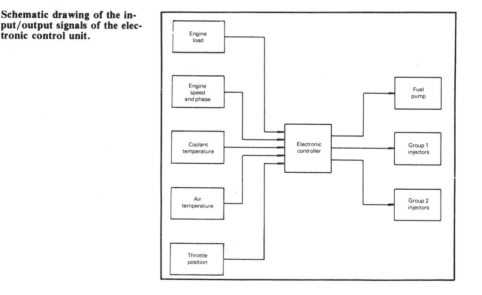

By using eight groups of five lines each, and a switching device that supplies base current to five metal-oxide-silicon transistors, the appropriate groups of lines can be connected so as to produce five-bit output words.

These digital output words are modulated by an interpolation function, and modified numbers representing load and speed are then sent out to the memory where they select the site which stores the fuel requirements for that particular combination of engine-operating conditions.

Lucas claims to have been first in the world to announce a digital-control fuel management system. Key to the Lucas invention is the large-scale integrated circuit, smaller than a postage stamp and yet containing over 3,000 components.

The fuel schedule is stored as a function of engine speed and load in the read-only memory, which has a capacity of 1,024 bits. It has sites to accommodate sixteen discrete values of speed and eight values of load.

What happens next? The memory output number, which spells fuel quantity, is fed into a number-to-time counter where it is stepped to zero by a device known as the fuel-trim oscillator. It is a metering unit. The time taken for the countdown to reach zero is proportional to the output number from the memory, and that time determines the opening duration for the injectors.

The output number from the memory does not go unchanged through the metering unit, but is modified by signals of air and coolant temperatures, plus idle, coasting and acceleration signals, which cause the frequency of the fuel-trim oscillator to rise or drop in response to diminished or increased fuel demand.

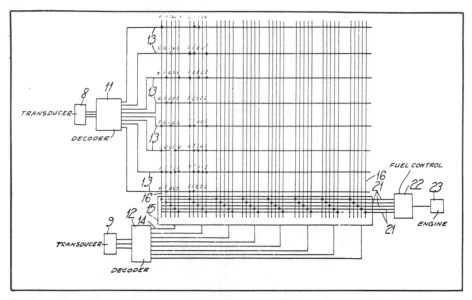

Diode matrix chooses two parameters out of three on which a continuous reading is fed into the control unit. The selection is constantly reviewed according to conditions. Output data comes in code of two-bit (or more) 'words' which then program the fuel metering.

Basic fuel-requirement map for all speed and load combinations are stored in the computer memory. The amount of fuel metered is proportional to injection time. Manifold pressure is read as "load."

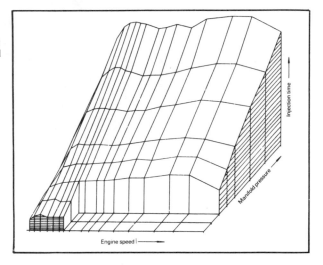

When these signals reach the fuel-trim oscillator, they are in digital form, while the transducers transmitted only analog data to the printed circuit.

The analog/digital conversion from the transducers takes place in several steps. When a signal is received from a transducer, it goes through a resistor, which feeds it to an amplifier. From there it enters the printed circuit, and if the signal is stronger than the minimum amplitude set by the circuit, an output is produced.

The output turns on a gate, which also receives an input from a clock source connected to a reversible binary counter. That starts operation in the binary counter, which turns on three transistors in succession. Each transistor is connected with the input line to the amplifier where the signal was treated before being admitted to the printed circuit.

The resistor and the binary counter are connected so that for a given analog voltage fed through the resistor, the binary counter will assume a stable condition in which one or more of the transistors is conducting, and the gate is turned off. By this method, the stable condition of the counter gives a digital signal which represents the analog voltage.

The injectors are operated in groups by solenoid valves. The solenoids are energized through power circuits referenced to the countdown period of the number-to-time counter. Since the pressure differential between the fuel line and the intake manifold is constant, the quantity of fuel injected is dependent only on the length of the injection period. A separate cold-start injector provides mixture enrichment for cold-start conditions.

Manifold pressure is measured by an inductive sensor whose action modifies the period of a timing circuit that sends pulses to a counter. A pair of read switches in the ignition distributor has a double function: The time between one closing of the switch and the next closing gives an accurate rpm reading, and each closing triggers the fuel-timing circuit for alternate groups of cylinders.

Lucas has adopted direct measurement of mass airflow for this system, with an airflow meter that feeds signals directly to the electronic control unit.

The electronic control unit operates from twelve-volt battery current. It has built-in compensation for normal voltage variations in the form of an extra pulse circuit that can alter the pulse length if battery output declines or rises. Electrical connections to the control unit are made with a multipole plug.

Chrysler's Single-Point Injection

After their Electrojector experience, the engineers at Chrysler waited a long time before again tackling the problems of developing a simple, reliable and low-cost electronic injection system.

Chrysler had been a leader, by virtue of its Aerospace Division in New Orleans and its Electronics Division in Huntsville, Alabama, in applying electronics to Chrysler-built cars, most notably by the standardization of electronic ignition on all car engines in 1972.

It was not until 1977, however, when Chrysler had arrived at the basic concept for a single-point continuous-flow injection system with electronic control, that a full-scale research program was started.

"We built twenty cars with electronic fuel injection in 1977, twenty more in 1978 and sixty-five in 1980," reported E. W. Meyer, Chrysler's chief engineer for motor/electrical. "We had to make sure we ironed out all the problems before we put the first system in a production car. We've tested the system under actual driving conditions at the Chelsea Proving Grounds, down in Arizona for hot-weather conditions, and in northern Ontario for extreme cold. In all, we have put over a million miles on these cars, and the results have been excellent—in reduced emissions, good fuel economy and outstanding driveability."

The Chrysler system was made standard on the 5.2-liter V-8 of the 1981 Imperial as the first step toward across-the-board availability.

The system offers complete electronic engine control, the same control module controlling the spark advance. It monitors the air/fuel ratio electronically, compares it to an ideal ratio and adjusts it automatically to changing environmental and engine conditions. Chrysler holds or has applied for twenty-four separate patents covering almost every part of the control system.

The most obvious innovation in the Chrysler system is the electronic metering of mass airflow as well as fuel quantity. In addition to giving higher precision, electronic measurement minimizes the effects of manufacturing tolerances and wear on the system's mechanical components.

The Chrysler system maintains the quality of the air-fuel mixture by arranging the throttle blades and bore in geometrical relationships that make the inducted air shear, entrain and distribute the fuel evenly to each cylinder. In addition, it electronically estab-

The throttle body assembly is centered in the air cleaner, which carries the electronic control unit.

lishes the base air/fuel ratio for each individual car. There is no need for a manifold absolute pressure sensor.

Chrysler's system consists of three major assemblies—each one a functionally complete unit that can be tested separately. The fuel-supply assembly is located inside the fuel tank. In addition to the conventional equipment that delivers fuel to the engine, this system also has an electric turbine pump and several check valves.

The second major assembly is the air cleaner, which also contains the airflow sensor, and the metering and ignition electronic module. Third comes the throttle body and mixture-control assembly. This includes the control pump and its power electronics, the fuel flow sensor, the pressure-regulating valves, the spray bars and the automatic idle-speed motor.

The computer receives input data on three separate functions: the flow of air into the engine, the flow of fuel and the oxygen content in the exhaust gas. It compares these signals to an ideal calibration. When any of the signals is different from the calibration, the computer signals the control-pump motor to deliver more or less fuel, depending on whether the mixture is too rich or too lean.

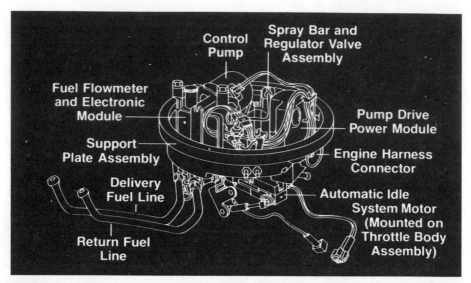

The fuel control subsystem rides on top of the throttle body and controls fuel metering on commands from the Combustion Computer.

The air-induction system receives air from two sources. First, unheated air from the fresh-air inlet. Second, heated air from the heat stove on the exhaust manifold. The two are mixed in proportions controlled by a vacuum actuator which moves a damper, admitting a greater or smaller amount of heated air.

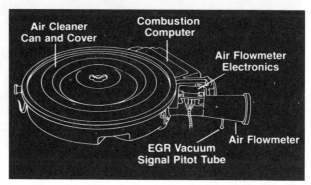

The pump in the tank feeds the fuel to the control pump, which takes a small portion of it, and delivers the fuel at a pressure of 21 psi at idle through the fuel flow sensor and the lower-pressure regulator valve, into the spraybar, and then to the nozzles. The control pump is a positive-displacement, slipper-type pump driven by a variable-speed direct-current motor.

Since the size of the openings for fuel is fixed, the pressure must be varied to deliver more fuel at higher speeds. At maximum speeds, the control pump delivers fuel at pressures up to 60 psi. The spraybar is designed to produce an even flow at low speeds. At higher speeds, a second spraybar automatically opens to deliver the full amount of fuel at the correct pressure. Fuel that is not needed to maintain a given speed is automatically returned to the tank through a low-pressure regulator valve and a return line.

The nozzles are connected to both the light-load and power spraybars. A number of airfoil-shaped nozzles are arranged in a circle around the injector body to refine the light-load fuel-spray pattern. The light-load circuit supplies all the fuel the engine needs up to a metering pressure of 34 psi.

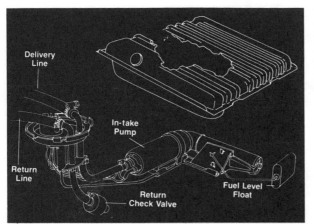

The fuel supply subsystem has an open circuit to the tank, returning all excess fuel to the tank from the control pump. The delivery pump is located inside the tank. Note: In-take pump is misprint for In-tank pump.

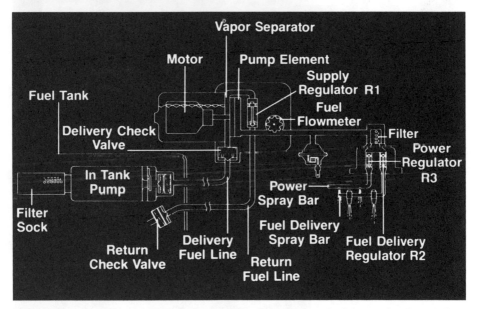

Fuel-supply system schematic. The flow meter has a free-spinning wheel, its rotational rate being proportional to fuel flow. The turning vanes interrupt the light path between a light-emitting diode and a phototransistor, and signals on flow rate are processed by the flow meter module for passing on to the Combustion Computer.

At higher pressure, the power circuit goes into action, while the light-load circuit continues to spray at maximum. The power injector bar adds its own spray from a single larger orifice.

The airflow meter is a completely new type. It consists of a large-diameter venturi, mounted in close proximity to the air-cleaner outlet, upstream from the throttle.

A ring of fixed, radially arranged vanes near the mouth of the venturi deflects the air into a clockwise swirl pattern. Due to the centrifugal effects of this swirling air, the air pressure is lower at the center of the vortex than at the outside.

This vortex looks like a miniature tornado whose eye looks like a finely coiled thread stretching throughout the length of the duct. Nearest the inlet end, the eye is cen-

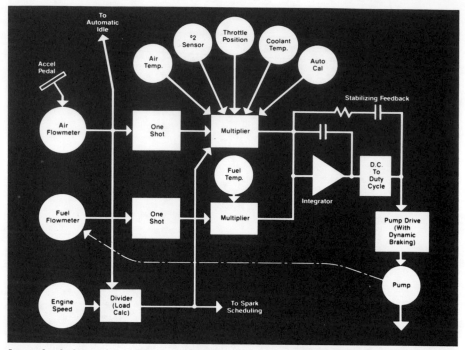

Inputs for fuel metering are shown in this schematic. The control pump is a positive-displacement, slipper-type pump driven by a variable-speed electric motor. The control unit controls air/fuel ratio by ordering changes in the pump speed.

Airflow sensor works by having a U-shaped probe keep track of the movement of the low-pressure 'eye' in the vortex.

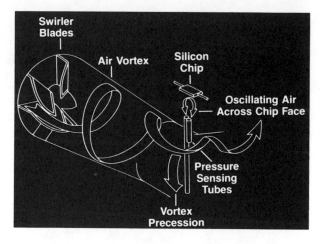

tered in the duct; but as it approaches the outlet, the vortex expands to fill the widening duct, becomes unstable, and the eye goes into an orbit around the center. What matters about this displacement of the eye is not the distance from the center or to the walls of the duct, but the frequency of its orbit.

This frequency is proportional to the volume of air passing through the duct per unit of time. Measuring the frequency of pressure variations along the orbit, therefore, provides an analog reading for mass airflow.

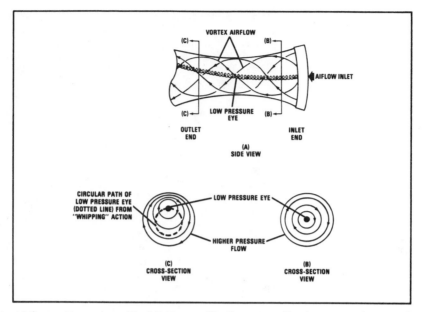

Vortex airflow patterns vary with airflow mass. The frequency of 'eye' movement is proportional to the volume of air passing through the duct in a given time.

This measurement is taken by a row of U-shaped pressure-probe standpipes located near the outlet from the venturi. A silicon chip converts the low-pressure differential readings into a high-powered electrical signal. As the volume of air increases, the signal pulses faster. The pulse frequency tells the computer how much air goes into the engine at any given instant.

The fuel flow meter consists of a cylindrical cavity containing a free-turning wheel resembling a gear, but its teeth are actually small, curved vanes. It works like a paddle wheel. Fuel enters the cavity on a tangent and its pressure is brought to bear against the vanes. This leads to rotation in the wheel at a rate that is proportional to the fuel flow. The greater the flow, the faster the wheel spins.

On one side of the paddle wheel is a light-emitting diode, and on the other side is a photo-transistor. As the paddle wheel spins, it interrupts the flow of light between the two and sets up a signal which pulses faster as more fuel flows by. These pulses are relayed to the computer, and their frequency tells the computer how fast the fuel is flowing.

To the engine, the information that really matters is not the volume of air or fuel flow. It's the air/fuel ratio that counts, not the volumetric ratio. Therefore, the system incorporates three additional sensors which translate the volume readings of air and fuel to mass readings.

The use of the term 'ideal' in connection with the air/fuel ratio earlier in this chapter must not be taken to mean that Chrysler's fuel-injection system is restricted to operation with an air/fuel ratio that does not vary from the stoichiometric.

E. W. Meyer explains: "We designed this system to operate on a very lean mixture of fuel and air—as little as one part fuel to twenty-one parts air. To operate the engine on a mixture this lean, you need three things: first, a uniform distribution of air and fuel to each cylinder; second, an even supply of air and fuel to a cylinder from cycle to cycle; and third, fuel preparation which gives the lowest specific fuel consumption and minimum specific oxides of nitrogen emissions, while at the same time controlling specific hydrocarbons output."

Fuel mixing takes place in an area starting upstream of the throttle body, where the fuel bars emerge from the low-flow and high-flow control valves, and reaching through the manifold to the intake valves.

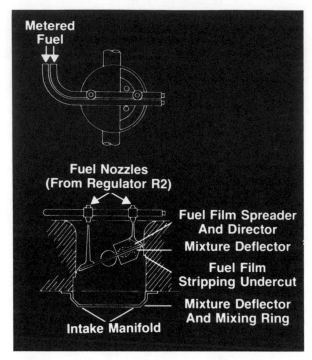

Metered Fuel

Fuel Nozzles
(From Regulator R2)

Fuel Film Spreader
And Director

Mixture Deflector

Fuel Film
Stripping Undercut

Mixture Deflector
And Mixing Ring

Intake Manifold

Basically, the signals from the two flow meters (air and fuel) trigger one-shot pulses of opposite polarities—positive and negative. These signals are fed into an integrator, which controls the pump motor.

An automatic calibration system individually establishes the air/fuel ratio on each car and compensates for changes in barometric pressure. There is a temperature sensor on the air-cleaner housing so the combustion computer can adjust for changes in temperature that affect the density of the air. An oxygen sensor in the exhaust system tells the computer when the air-fuel mixture is too rich or too lean.

The fuel flow also makes contact with a silicon thermistor which senses the temperature of the fuel. Its electrical conductivity varies in proportion with the fuel temperature, and it is thus able to send fuel-temperature signals to the computer.

The mixture-control device is a conventional two-part, single-throttle-shaft design like the air horn on a carburetor—with several important differences:
1. All the fuel is introduced above the throttle blade.
2. There are two fuel-distribution bars over each port—a primary bar for low-fuel flow, and a power bar for staged, high fuel flow.
3. The conventional butterfly throttle blade has a fuel-film spreader mounted above the blade and a mixture deflector mounted below.

The fuel streams impinge against the film spreader, which distributes the gasoline outward into thin films. The lower-velocity air at the larger blade openings draws the fuel film toward the sharp edges of the blade, resulting in a highly atomized mixture of fuel and air. The lower mixture deflector improves the cylinder-to-cylinder mixture uniformity.

A high degree of fuel atomization is vital for efficient and uniform combustion. A spherical droplet of fuel has a surface area proportional to the square of the diameter and a volume proportional to the cube of the diameter. Smaller and smaller atomized fuel droplets are vaporized in less time for an equal amount of heat.

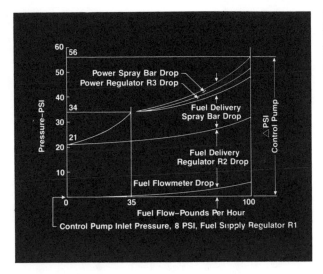

Fuel flow is related to control-pump pressure in this chart, with curves showing the pressure drop occurring in the fuel flow meter, regulators and spraybars.

Finely atomized fuel droplets of ten to fifty microns seem to be the right size. They mix more easily with the intake air and remain in the manifold longer. The action of the air shearing the fuel off sharp edges seems to be a very simple and effective device to achieve these droplet sizes.

A fuel-pressure switch is included in the fuel-control circuit between the flow meter and the injection assembly. Normally it is open, which indicates that the system has sufficient pressure for starting the engine. The valve is closed when the pressure is too low, and its closing completes a bypass circuit that drives the control pump at full speed until adequate pressure is restored.

When the engine is cranking, idling or running under a light load, only the primary fuel-bar nozzles are delivering fuel. As speed increases, and the fuel flow reaches about thirty-five pounds per hour, the power-regulator valve opens and fuel starts to flow from the power orifices.

The computer is a common center for four separate electronic circuits: The first is the fuel-injection circuit which gives speed orders to the control pump and thereby provides the basic factor for control of the air/fuel ratio. The second circuit is the automatic calibration circuit that monitors the oxygen feedback from the exhaust gas and serves to fine-tune the first circuit. The third is the electronic spark-advance circuit that provides the initial ignition current and spark timing. The fourth is the automatic idle-speed circuit which goes into action whenever the throttle is closed, and stabilizes the idle.

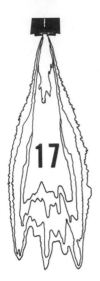

Ford's Electronic Engine Control System

Except for Ford's Indy-type (with Bendix) and the Cosworth-Ford Grand Prix racing engines (with Lucas), the Dearborn auto maker had long shied away from fuel injection of any kind. All its US-built production cars remained faithful to the carburetor until 1980, and in Europe only a few special models were equipped with K-Jetronic.

Ford introduced for certain of its 1978 models the electronic control of engines and an oxygen sensor which could 'feed back' signals to an electrically controlled carburetor. For some 1979 models, the two concepts were married into a second-generation system (EEC-II). For 1980, a third generation of engine electronics (EEC-III) was introduced.

The EEC-III system was combined with an electronically operated carburetor that can just as well be described as an electronically controlled single-point pulse-time-modulated fuel-injection system. It was introduced on the 1980 Lincoln Continental and Continental Mark VI powered by the five-liter (302-cubic-inch) V-8 engine, matched with Ford's Automatic Overdrive transmission. Ford claimed that this combination in the newly designed cars, having smaller outside dimensions and lighter weight, produced exceptional fuel-economy figures—25 mpg for the highway estimate and 17 mpg for the EPA estimate for city driving.

For 1981, use of EEC-III with electronic fuel injection was extended to the Ford LTD and Mercury Marquis.

In some applications the EEC-III will be replaced in 1984 by EEC-IV, a more versatile system with simplified componentry. It acts several times faster in its response to altered conditions, and has twenty percent greater memory capacity despite two-thirds fewer large-scale integrated circuit chips.

A complete description should start at the beginning, so we must go back to 1978 and the EEC-I. It was a completely solid-state module using a digital microprocessor and other custom-designed integrated circuits. Seven sensors were used to determine crankshaft position, throttle position, coolant temperature, inlet air temperature, manifold absolute pressure, barometric pressure and exhaust gas valve position. Using this information, the module calculated spark advance and the exhaust gas recirculation flow rate.

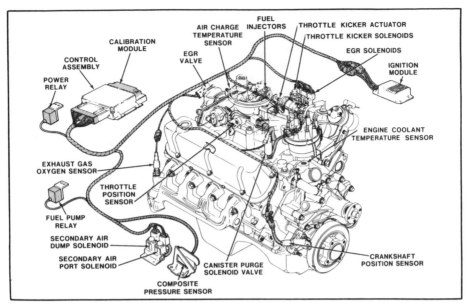

Ford's electronic fuel-injection system is a single-point type with two injectors mounted in the throttle body. A microprocessor controls fuel metering, spark timing and various other engine functions.

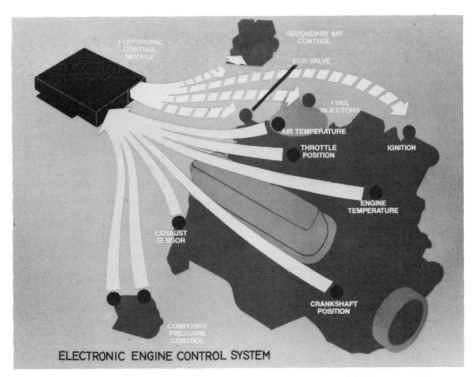

ELECTRONIC ENGINE CONTROL SYSTEM

Ford's electronic engine control system gets input data on engine speed (crankshaft position), throttle position, coolant temperature, charge air temperature and oxygen level in the exhaust gas. The microprocessor controls air/fuel ratio, spark timing and exhaust-gas recirculation.

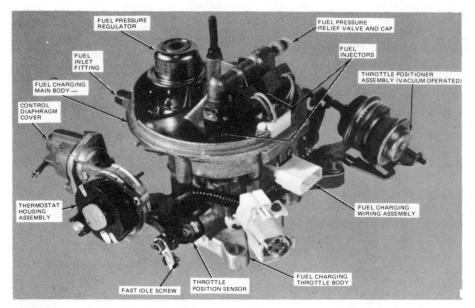

Injector valves are solenoid-operated, the amount of fuel injected varying only with valve opening duration, as ordered by the microprocessor.

Ford's injector uses a pintle-type nozzle delivering a fine fuel spray into the throttle passage. Fuel is fed under constant pressure, and the injection is timed electrically. Solenoid chamber is sealed off by O-rings. Pintle nozzle seats against lower end of nozzle holder. Quantity depends on opening duration.

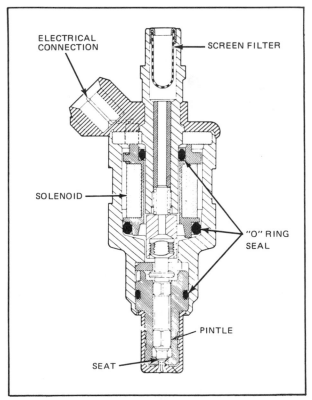

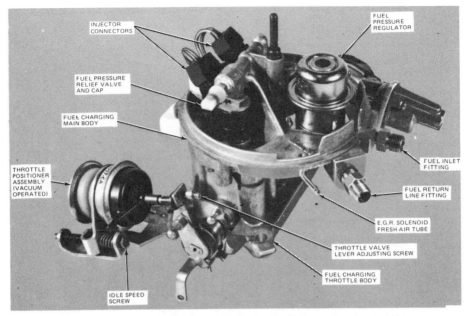

INJECTOR CONNECTORS

FUEL PRESSURE REGULATOR

FUEL PRESSURE RELIEF VALVE AND CAP

FUEL CHARGING MAIN BODY

THROTTLE POSITIONER ASSEMBLY (VACUUM OPERATED)

FUEL INLET FITTING

FUEL RETURN LINE FITTING

E.G.R. SOLENOID FRESH AIR TUBE

THROTTLE VALVE LEVER ADJUSTING SCREW

FUEL CHARGING THROTTLE BODY

IDLE SPEED SCREW

Mechanically, Ford's electronic fuel injection is far simpler than a carburetor, but it has numerous electrical connections that are not used on conventional carburetors. Throttle position is set by vacuum—not by mechanical linkage.

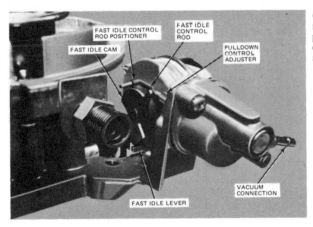

FAST IDLE CONTROL ROD POSITIONER

FAST IDLE CONTROL ROD

FAST IDLE CAM

PULLDOWN CONTROL ADJUSTER

VACUUM CONNECTION

FAST IDLE LEVER

Closeup view of fast-idle mechanism. Control diaphragm is located inside its cover (with vacuum connection at the end).

The next step was to combine the EEC-I with a three-way catalytic converter using oxygen feedback control. It was known as EEC-II, and was less complex, more reliable and lighter than the separate systems used earlier.

The first step in the development of the EEC-II's computer circuit was created by the engineers in Ford's EEC department to define the system requirements. The technical specifications were turned over to its semiconductor suppliers for integrated circuit design and fabrication. Through a complex process, the circuit was reduced to its final size— a chip about one-quarter-inch square. The EEC-II system had six chips, and each one contained 10,000 to 15,000 electronic devices.

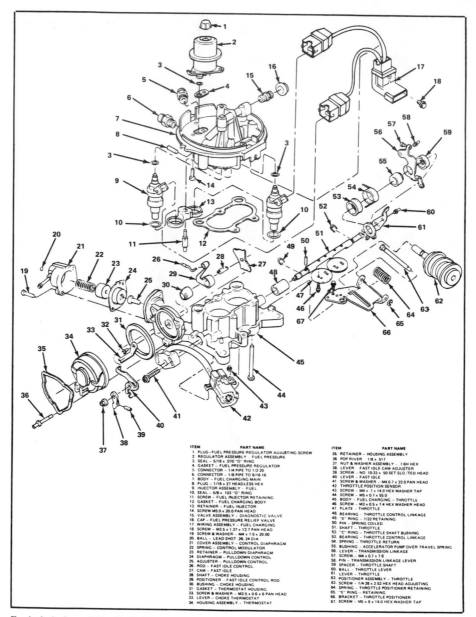

ITEM	PART NAME	ITEM	PART NAME
1. PLUG - FUEL PRESSURE REGULATOR ADJUSTING SCREW		35. RETAINER - HOUSING ASSEMBLY	
2. REGULATOR ASSEMBLY - FUEL PRESSURE		36. POP RIVER - 1/8 x 517	
3. SEAL - 5/16 x .070 "O" RING		37. NUT & WASHER ASSEMBLY - 7-6H HEX	
4. GASKET - FUEL PRESSURE REGULATOR		38. LEVER - FAST IDLE CAM ADJUSTER	
5. CONNECTOR - 1/4 PIPE TO 1/2 20		39. SCREW - NO 10-32 x 50 SET SLO.TED HEAD	
6. CONNECTOR - 1/8 PIPE TO 9/16-16		40. LEVER - FAST IDLE	
7. BODY - FUEL CHARGING MAIN		41. INJECTOR ASSEMBLY - FUEL	
8. PLUG - 1/16 x 27 HEADLESS HEX		42. THROTTLE POSITION SENSOR	
9. INJECTOR ASSEMBLY - FUEL		43. SCREW - M4 x 7 x 14.0 HEX WASHER TAP	
10. SEAL - 5/8 x .103 "O" RING		44. SCREW - M5 x 0.7 x 55.0	
11. SCREW - FUEL INJECTOR RETAINING		45. BODY - FUEL CHARGING - THROTTLE	
12. GASKET - FUEL CHARGING BODY		46. SCREW - M3 x 0.5 x 7.4 HEX WASHER HEAD	
13. RETAINER - FUEL INJECTOR		47. PLATE - THROTTLE	
14. SCREW M5.0 x 20.0 PAN HEAD		48. BEARING - THROTTLE CONTROL LINKAGE	
15. VALVE ASSEMBLY - DIAGNOSTIC VALVE		49. "E" RING - 7/32 RETAINING	
16. CAP - FUEL PRESSURE RELIEF VALVE		50. PIN - SPRING COILED	
17. WIRING ASSEMBLY - FUEL CHARGING		51. SHAFT - THROTTLE	
18. SCREW - M3.5 x 1.27 x 12.7 PAN HEAD		52. "C" RING - THROTTLE SHAFT BUSHING	
19. SCREW & WASHER - M4 x 7.0 x 20.00		53. BEARING - THROTTLE CONTROL LINKAGE	
20. BALL - LEAD SHOT .26-.24 DIA		54. SPRING - THROTTLE RETURN	
21. COVER ASSEMBLY - CONTROL DIAPHRAGM		55. BUSHING - ACCELERATOR PUMP OVER TRAVEL SPRING	
22. SPRING - CONTROL MODULATOR		56. LEVER - TRANSMISSION LINKAGE	
23. RETAINER - PULLDOWN DIAPHRAGM		57. SCREW - M4 x 0.7 x 7.6	
24. DIAPHRAGM - PULLDOWN CONTROL		58. PIN - TRANSMISSION LINKAGE LEVER	
25. ADJUSTER - PULLDOWN CONTROL		59. SPACER - THROTTLE SHAFT	
26. ROD - FAST IDLE CONTROL		60. BALL - THROTTLE LEVER	
27. CAM - FAST IDLE		61. LEVER - THROTTLE	
28. SHAFT - CHOKE HOUSING		62. POSITIONER ASSEMBLY - THROTTLE	
29. POSITIONER - FAST IDLE CONTROL ROD		63. SCREW - 1/4-28 x 2.53 HEX HEAD ADJUSTING	
30. BUSHING - CHOKE HOUSING		64. SPRING - THROTTLE POSITIONER RETAINING	
31. GASKET - THERMOSTAT HOUSING		65. "E" RING - RETAINING	
32. SCREW & WASHER - M3.5 x 0.6 x 6 PAN HEAD		66. BRACKET - THROTTLE POSITIONER	
33. LEVER - CHOKE THERMOSTAT		67. SCREW - M5 x 8 x 14.0 HEX WASHER TAP	
34. HOUSING ASSEMBLY - THERMOSTAT			

Exploded view of Ford's fuel-charging assembly. The number of parts is about the same as in a two-barrel carburetor; the big difference being the lack of a float bowl.

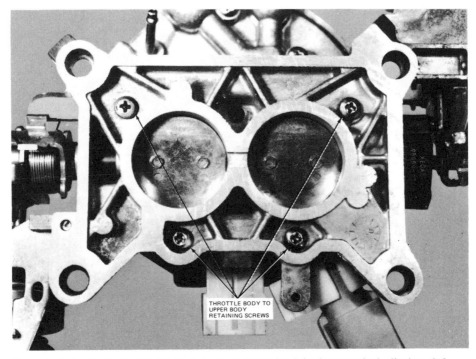

THROTTLE BODY TO
UPPER BODY
RETAINING SCREWS

Upper body is attached to throttle body by two screws. The joint between the bodies is sealed by a gasket.

Henry A. Nichol, then Ford's chief power train engineer, explained what was achieved in going from the first to the second generation of EEC systems: "We added capabilities to the system, but were able to reduce the package size by forty percent, the number of computer parts by thirty percent, and the weight by almost fifty-five percent. The system complexity was reduced and the costs were lowered dramatically. A third generation of the EEC system, to be introduced on 1980 cars, will be even less complex, but will provide more control functions."

In addition to controlling the major engine functions, EEC-II provided many other benefits. For one, it controlled the purging of vapors in the storage canister used with the fuel-evaporative emissions-control system. Under certain conditions, the purging caused the carburetor to run rich. When that happened, however, the EEC-II system simply sensed this change through the exhaust gas sensor and signaled the proper adjustment to the carburetor.

Another benefit was minimizing variations in idle speed through a throttle-idle positioner. For example, a sensor indicated whether the vehicle's air conditioning system was operating. If it was, the control module signaled a throttle-position solenoid to increase the throttle opening, compensating for the increased engine load while maintaining an acceptable idle speed. If the air conditioner was not being operated, the throttle-idle opening was reduced and thereby saved fuel. The throttle-idle positioner made similar corrections for cold-engine conditions, and for high altitude.

The EEC-II system eliminated the need for special calibrations in high-altitude areas. In addition to the idle-speed adjustment, the system sensed barometric pressure and made the necessary engine adjustments automatically.

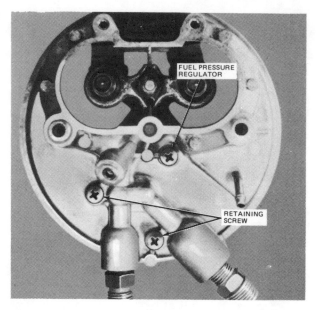

Fuel pressure regulator can be removed after undoing three screws. The fuel-charging wiring assembly should first be disconnected.

FUEL PRESSURE REGULATOR

RETAINING SCREW

The system also controlled the thermactor—a pump that adds air to the exhaust system—by diverting this secondary air to either the catalyst or the exhaust manifold as required to maximize emissions control.

EEC-II controlled idle mixture and initial spark automatically, thus eliminating the need to make these maintenance adjustments.

The Model 7200 'feedback' carburetor in the EEC-II system was the same Variable-Venturi carburetor introduced on the 5.0-liter (302-cubic-inch) V-8 engines in California in 1977 except for the modifications that were necessary to include the feedback feature. One major modification was the addition of a stepper motor that adjusted the position of a rod in the carburetor to trim the airflow—rich or lean—required to maintain the desired ratio.

The essential difference between EEC-III and EEC-II lies in the use of single-point electronic fuel injection on the newer system. The control system, its sensors and outputs, are basically the same.

Ford's electronic carburetor used with EEC-III is produced at the Rawsonville, Michigan, plant of the Electrical and Electronics Division, where most of the company's conventional carburetors also are manufactured.

The microprocessors for the EEC-II system were supplied by Shibaura Electric Company (Toshiba) and the Electrical and Electronics Division of Ford Motor Company. Microprocessors for the EEC-III system are produced by Motorola. National Semiconductor Company and Signetics supplied small integrated circuits for EEC-II, while suppliers for the EEC-III included Intel, producing large-scale integrated circuits; National Semiconductor Company, producing small-scale integrated circuits and transistors; and Fairchild Semiconductor Company, producing small-scale integrated circuits.

At the heart of the electronic fuel-injection system, Ford placed two electrically activated fuel metering valves. When energized, these valves spray a precise quantity of fuel into the engine's intake airstream. Both injector nozzles are mounted vertically above the throttle plates and connected in parallel with the fuel pressure regulator. The injector valve bodies consist of a solenoid-actuated pintle and needle valve assembly. An electric control signal from the EEC-III electronic processor activates the solenoid causing the pintle to move inward off its seat and allows fuel to flow. The injector flow orifice is fixed,

RETAINING SCREW

Both injectors are clamped to the upper body, with easy removal after undoing one central retaining screw. The injectors should be marked for throttle-side or choke-side installation.

and the fuel supply pressure is constant, therefore, fuel flow to the engine is controlled by how long the solenoid is energized.

The controlled, high fuel pressure in the injectors, along with the precise change in fuel volume determined by EEC-III, improves fuel distribution to each cylinder, compared with the carburetor it replaces. This in turn provides optimum driveability and economy while exhaust emissions are maintained at acceptable levels.

Fuel to the injectors is provided by a high-pressure electric pump inside the fuel tank. A special primary fuel filter is located in the fuel line, under the passenger compartment, along with a smaller, secondary filter in the engine compartment.

A fuel pressure regulator mounted to the carburetor, or throttle body, just ahead of the injectors, controls fuel pressure to a precise and constant value of 39 psi. Excess fuel supplied from the pump but not needed by the engine is returned to the fuel tank by way of a fuel-return line.

The pressure regulator is mounted on the fuel-charging main body near the rear of the air horn surface. The regulator is located so as to nullify the effects of the supply-line pressure drops. It is designed so that it is not sensitive to back pressure in the return line to the tank.

A second function of the pressure regulator is to maintain fuel supply pressure following engine and fuel pump shutdown. The regulator functions as a downstream check valve and traps the fuel between itself and the fuel pump. The maintenance of fuel pressure after engine shutdown prevents fuel-line vapor formation and allows rapid re-starts and stable idle operation.

Airflow to the engine is controlled by two butterfly valves mounted in a two-piece, die-cast aluminum housing called the throttle body. The butterfly valves are identical in configuration to the throttle plates of a carburetor and are actuated by a similar linkage and pedal cable arrangement.

A throttle position sensor is mounted on the throttle shaft on the side of the fuel-charging assembly and is used to supply a voltage output proportional to the change in the throttle plate position. This sensor is used by the computer (EEC-III) to determine the operation mode (closed throttle, part throttle, wide-open throttle) for selection of the proper fuel mixture, spark and EGR at selected driving modes.

For fast idle before the engine reaches normal operating temperature, Ford uses a throttle-stop cam positioner of the same type that is used on carburetors. The cam is positioned by a bimetal spring and an electric positive temperature heating element. The electrical source for the heating element is 7.3 volts from the alternator stator, which provides voltage only when the engine is running.

The heating element is designed to provide the necessary warmup profile in accordance with the starting temperature (cold engine prior to cranking) and the length of time after starting. Multiple positions on the cam profile allow for a gradual slowdown from cold-engine speed to curb idle speed during warmup. A second feature of the cold-engine speed control is automatic kickdown from high cam (fast idle) engine speed to some intermediate speed. This is accomplished by the computer (EEC-III) vacuum signal to the automatic kickdown motor which physically moves the high speed cam a predetermined time after the engine starts.

The digital-interactive control feature in the system was developed and patented by Ford.

Ford admits that it is partly because of cost considerations that its electronic engine control with fuel injection is used only on a selected engine in a few expensive models.

"When you consider the man-hours and development expense to produce a system as complex as this, there is bound to be a high cost factor," Henry A. Nichol explained. "It was no different with this program. There were extremely large startup costs, but our experience to date has confirmed that in the long run, this system is going to be affordable for both our customers and us.

"For our customers, it will mean lower operating costs, higher reliability and better performance. For us, it will mean more efficient production, better quality and a more efficient way to meet emissions and fuel-economy standards."

Further explanation came from John A. Betti, Ford vice president for power train and chassis operations, who told the press in July 1980: "We know how and where to make the most cost-effective use of engine electronics and don't feel the need to apply them across the board. We know and understand what electronic control systems can do for us, and for that reason we can pick our shots without underutilizing a powerful tool or adding unnecessary cost to our products.

"Our strategy is to examine each engine and power train combination to determine the most effective application. That may mean using interactive controls on some engines, stand-alone systems on others or conventional systems on some. One of the side effects of our pioneering work in this field is that it has given us a better understanding of how the internal-combustion engine works, and has shown us how to achieve some gains through non-electronic means.

"For example, during the development of some of our 1981 engines, we started out with electronic engine controls, but found ways to achieve our fuel-economy and emissions goals without them."

150

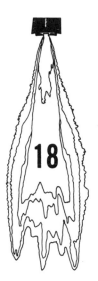

18

Zenith Fuel-
Management Systems

Zenith is the trademark chosen by Pierburg (formerly Deutsche Vergaser Gesellschaft) for its alternative fuel-management systems. Alternative, that is, to the carburetor, which is Pierburg's main line of business. There is not one Zenith system, but five. Three have evolved sequentially, and two remain separate but parallel. We will deal with them in chronological order.

The first Zenith system of which details were released appeared in 1973 and was identified by the letter C. Two preceding systems, A and B, never made it through the experimental stage.

It was probably natural for a carburetor manufacturer to devise an injection system that works with uncommonly low pressure, in which atomization starts within the injector nozzle and provides continuous injection.

The company's vast experience with carburetors of all types has obviously been influential in the choice of mechanical linkages and hydraulic and pneumatic connections where a fuel-injection expert would use other means (electrical or electronic).

On the Zenith C, only the fuel feed pump was electrical. This pump drew fuel from the tank and pushed it through a filter on its way to the pressure regulator. The pressure regulator was a valve which leveled-off the line pressure to a constant value of 30 to 45 psi. Its output line led straight to the metering distributing unit.

This unit was the heart of the whole system, mixing fuel with air in accordance with two separate measurement devices, and delivering a uniform mixture to an individual injector for each cylinder intake port.

The translation from input measurements to output commands was centered on an eccentric cone, keyed to its spindle for rotation, but free to move axially. The cone was displaced axially on the shaft in direct proportion to the movement of a diaphragm air pump driven from the engine camshaft. The air pressure from the pump served as a measure of engine speed (rpm), and was piped to a roll-sock type of diaphragm on one end of the cone spindle displacing the cone against the load of a coil spring along its own spindle. This spindle ran horizontally.

The air pressure was applied on the small end of the cone, and the spring load on the base end. A roller riding on the surface of the cone, but held in a fixed vertical arc by a pivoted lever, was, therefore, kept in a raised position by spring force; but it was allowed to descend to lower heights when the cone was moved by diaphragm force.

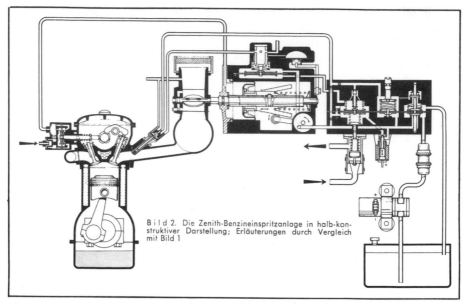

Bild 2. Die Zenith-Benzineinspritzanlage in halb-kon-
struktiver Darstellung; Erläuterungen durch Vergleich
mit Bild 1

The original Zenith system did not have airflow measurement. Fuel was pumped to the pressure
control unit, which was also connected to the fuel metering unit. A cone-and-follower arrangement gave mechanical adjustments in fuel quantity.

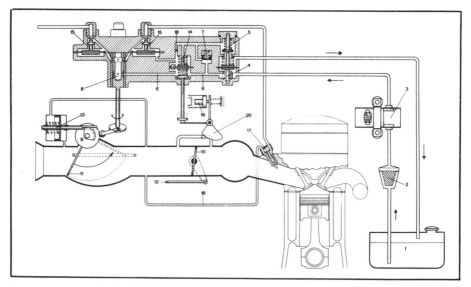

The eventual Zenith CL system embodied airflow measurement with pivoted flap (11), positioned
upstream of the throttle valve (10). 1-Tank, 2-Filter, 3-Pump, 4-Holding valve, 5-Return valve, 6-
Supply line, 7-Pressure valve, 8-Metering distributor, 9-Cone, 12-Throttle linkage, 13-Amplifier,
14-Control pressure valve, 15-Differential pressure valve, 16-Restrictor, 17-Injector, 18-Atom-
ization air line, 19-Expander element, 20-Supplementary air valve.

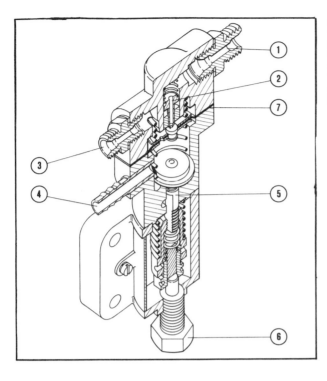

Control pressure valve of the CL system. 1-Return-line connection, 2-Valve seat, 3-System pressure adjustment screw, 4-Fuel inlet line, 5-Pushrod with spring-loaded plate, 6-Adjustment screw, 7-Diaphragm.

Because of the eccentricity of the cone, its rotation also affected the position of the roller.

The cone spindle was connected to the throttle plate via an adjustable, mechanical coupling so as to obtain a reading on throttle angle. Opening or closing the throttle plate produced rotation in the spindle and the keyed-on cone.

The idea of the cone may have been borrowed from Kugelfischer. It served the same purpose; namely, that of having each position of the roller on the surface of the cone correspond to a specific combination of engine speed and load. Consequently, both of these input measurements would be expressed in a single-plane motion, the vertical-arc described by the pivoted lever carrying the roller.

This lever conveyed the fuel-requirement message to a circular disc valve, mounted on the same pivot shaft. Near its periphery, the disc had an open axial channel. The disc valve housing was inserted in the fuel line, having an inlet port and an outlet port which registered with the channel in the disc. Disc rotation moved the channel relative to the ports so as to open or restrict the fuel flow through it.

From the disc (metering) valve, the fuel flow continued to the distribution section of the unit. It consisted of a base chamber with a diaphragm, and an upper chamber with a number of outlet ports corresponding to the number of the cylinders in the engine.

The fuel line entered the base chamber on the upper side of the diaphragm, which opened and closed a valve on the entry to the upper chamber. Counteracting the fuel-line pressure on the upper side of the diaphragm was another force, a secondary hydraulic pressure bled off from the pressure-control valve.

Diaphragm (and valve) movement was transmitted to a valve block inside the upper chamber, with an orifice for each outlet. Displacement of this valve block regulated the size of each outlet orifice. Consequently, the diaphragm worked as a differential-pressure valve, whose duty it was to equalize the fuel delivery to each injector.

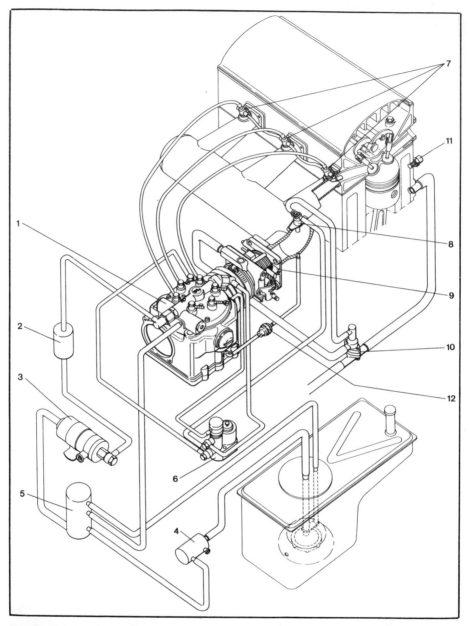

Schematic view of the DL system. Fuel is drawn from the tank by feed pump (4) and reaches the pressure pump (3) via an intermediate tank. Pressurized fuel reaches the control unit (1) via a final filter (2) and is distributed to the injectors (7). A cold-starting valve (8) provides fuel enrichment as needed, and a supplementary air valve (10) assures a stable idle. Pressure in the hydraulic control system is maintained by pressure valve (6). A valve (12) gives protection from backfiring. A thermo-time switch (11) limits warmup enrichment. (9) is the throttle body.

154

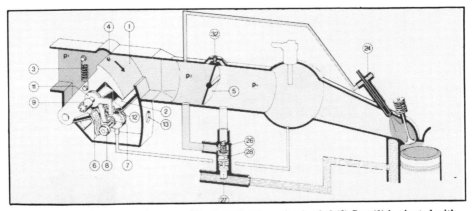

Details of the airflow measuring system in the DL. The spring-loaded (3) flap (1) is pivoted with a damper (2). Its movement is read by the roller (8) on the cam (6) and relayed via the lever (7) to the piston (9). The idle setting screw (4) blocks the throttle (5) from total closing. The supplementary air valve (26) is worked by a piston (28) according to the expansion of a heat-sensitive (27) element in the coolant flow.

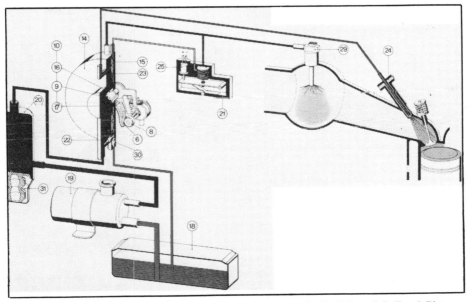

The intricacies of DL fuel metering are crowded into the control unit. 6-Cam, 8-Roller, 9-Piston, 10-Metering valve, 14-Metering distributor, 15-Differential pressure valve, 16-Sleeve, 17-Ring, 18-Tank, 19-Fuel pump, 20-Fuel filter, 21-Control pressure valve, 22-Fuel pressure control valve, 23-Diaphragm, 24-Injector, 30-Holding valve, 31-Reservoir.

Any change in the load on the engine produced a transitory change in the pressure drop across the disc valve. This immediately resulted in a displacement of the distributor diaphragm, so that constant pressure drop was reestablished.

The system included a form of acceleration pump, working by means of a cam on the cone spindle, offset from the cone base. It responded to opening of the throttle by causing a sudden increase in the fuel pressure in the chamber above the distributor diaphragm. That kept the valve open to let more fuel into the distributor valve.

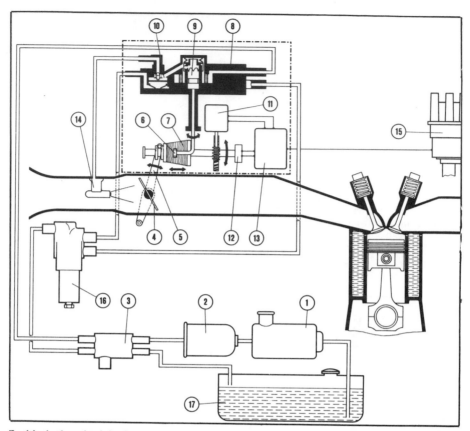

Zenith single-point injection system. One single injector is aimed into the throttle opening, metered by electronic control. 1-Electric fuel pump, 2-Fuel filter, 3-Pressure holding valve, 4-Throttle plate, 5-Axial movement of the cone, 6-Cone follower, 7-Cone, 8-Metering distributor, 9-Metering valve, 10-Differential pressure valve, 11-Servo motor, 12-Potentiometer, 13-Electronic control unit, 14-Injector, 15-Ignition distributor, 16-Control pressure regulator valve, 17-Tank.

The injector nozzle was positioned closely behind the inlet-valve head, mounted in a mantle, separated from the nozzle body by an open space over most of its length. The nozzle injected a very fine, low-pressure spray which began to atomize inside the injector mantle.

It is noteworthy that an open connection existed between the nozzle and the mouth of the throttle venturi. This passage served to supply atomization air into the space inside the mantle and had its greatest effect during conditions of high-pressure differentials, such as at idle and part-load operation.

The injector valves were closed when the engine was not running. When the engine was started, the fuel pressure overcame the spring load on the needle and a modulated resistance pressure provided by a hydraulic line from the pressure regulator.

The system also included a cold-start device consisting of a magnetic valve, governed by a thermo-time switch. It supplied additional fuel into the line behind the disc valve. The warmup valve assured the gradual lessening of mixture enrichment during warmup.

The closed space below the diaphragm in the lower chamber of the distribution section was pressurized under normal operating temperatures. This pressure acted to close the warmup valve in response to a thermostat submerged in the coolant. It reduced

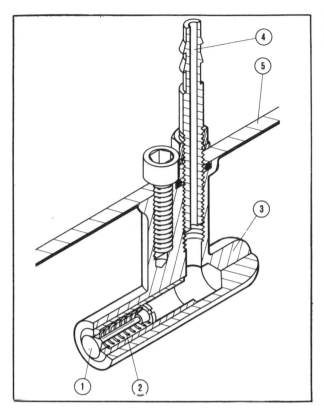

the pressure on the lower face of the diaphragm so as to admit more fuel until the engine reached normal operating temperature, while simultaneously turning the throttle plate so as to add sufficient air. This gave a fast-idle setting for as long as necessary.

A noreturn valve on the entry port to the pressure regulator made sure that the system remained pressurized when the ignition was turned off. That prevented evaporation and vapor lock, and facilitated hot-starts.

Zenith C gave excellent results under wide-open-throttle acceleration, but was not so satisfactory in a variety of transient conditions met in everyday driving. Consequently it tended to be used only in racing cars.

In an attempt to overcome these objections, Pierburg introduced the Zenith CL. The L stands for 'luft' (German for air), to indicate that this system used airflow metering. Here again, Pierburg did not copy Bosch but invented on its own.

However, Zenith CL also used a flap in the intake duct, positioned upstream of the throttle. The flap angle was determined by the pressure differential between the open area in front and the partial depression created in the section between the flap and the throttle plate. This pressure differential was measured across a small hole in the flap.

The flap was L-shaped in profile and pivoted on a spindle located below the duct. It was spring-loaded toward the closed position where a bulge on the duct provided a simple form of bypass for idle air, and so forth. An accordion bellows on the inboard side of the flap acted as a damper on flap movement during disruptions such as airflow fluctuations and inertia reactions.

Control pressure-regulating valve for the Zenith single-point injection system. It controls the pressure that balances the fuel delivery pressure across the diaphragm in the differential pressure valve. 1-Valve body, 2-Pushrod with diaphragm, 3-Spring, 4-Vent hole, 5-Spring base, 6- and 7-Differential pressure springs, 8-Valve plate, 9-Pressure rod, 10-Stroke transfer link, 11-Adjustment screw, 12-Heat-sensitive element, 13-Electrical terminal, 14-Manifold vacuum connection, 15-System pressure in, 16-System pressure out.

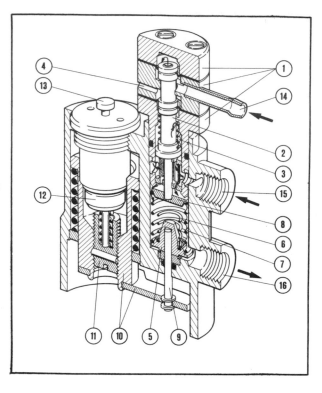

The movements of a three-dimensional cam mounted on the flap shaft were picked up by a pivoted lever with a roller at its end. This mechanism replaced the cone used in Zenith C. The three-dimensional cam reacted to changes in mass airflow by radial movement, pivoting on its shaft.

There was no longer any camshaft-driven air pump to give rpm readings by way of a diaphragm. Instead, there was a vacuum connection from the manifold, downstream of the throttle, direct to a diaphragm carried on the shaft of the roller running on the cam track. Movement of this diaphragm would displace the roller axially so as to move it to a different part of the cam surface.

The pivot arm that carried the roller transmitted its swinging motion directly to the piston in the metering unit. This was a new design based on the same principle of sliding ports.

The basic principle for the metering and distribution depended on control of the average pressure in the distributor unit, which consists of a piston, a sleeve and a ring. The piston was turned by the pivot arm, while the sleeve was held still.

The control piston had a number of triangular slots, one per cylinder in the engine. The sleeve surrounding the piston had similar slots, matching in size and number. Rotation of the piston inside the sleeve altered the cross-section of the slots that were open to let the fuel pass, and thereby regulated the amount of fuel supplied to the injectors.

In 1976 this system evolved into Zenith DL, receiving a separate starting-valve which sprayed fuel into a spherical bulge on the manifold, downstream from the throttle, whenever a thermostatically governed pressure valve relented enough to let its pressure drop below fuel-line pressure, which would open the line.

In 1977 Pierburg introduced an electronically controlled continuous-flow port-injection system for racing cars. An electrically driven pump draws fuel from the tank and pressurizes it to about 60 psi. After filtration, the fuel arrives at the metering unit and is distributed to the individual injectors.

Excess fuel is spilled off and returned through a system-pressure valve, a pressure-regulating valve and a depressurizing valve to the tank.

The metering unit contains a spill plunger that is lined to the main throttle and an electric motor actuated from the electronic control unit. Rpm signals reach the control unit from the ignition distributor. It also receives information on load from the throttle linkage. The injectors have pintle-type nozzles and receive a small amount of bypass air from the upstream side of the throttle to assure proper atomization during light-load operation.

Finally, there is the Zenith EL single-point injection system which, from the outside, looks confusingly like a carburetor.

Indeed, it has much in common with the air-valve carburetor, including an oil-damped piston that adjusts its position in a cylinder according to manifold vacuum.

The fuel metering is arranged by the same type of devices that are used in the Zenith DL. The nozzle has different dimensions but also includes a passage for atomization air to get inside the mantle.

Fiat/Marelli
Experimental System

Marelli is an electrical-equipment subsidiary of Fiat that has recently branched out into electronics. Marelli produced the electronic ignition system that was standard on the Fiat Dino in 1967-69. About 1973, Fiat and Marelli began to collaborate on an electronic fuel-injection system under the direction of R. Rinolfi of Fiat's research center. He courageously adopted an advanced concept, with a digital microprocessor and ultrasonic air-flow meter.

By 1977 the system had undergone its initial test phase, and details were released. The system remains experimental, however, as Fiat uses Bosch L-Jetronic on the 1981 two-liter 132. This does not mean, however, that Fiat and Marelli have given up on their own system.

Injectors were mounted at the intake ports, with timed delivery through electro-magnetic valves as directed from the control unit.

Fuel metering was calculated on the basis of signals from an engine speed sensor, engine stroke sensor, pressure transducer, coolant temperature transducer and the ultra-sonic airflow meter, which was the only truly unusual component in the system.

Fuel was taken from the tank by an electrical feed pump, pushed through a filter and fed to the pressure-regulating valve. Air entered the ultrasonic flow meter straight from the air cleaner and flowed from there to the throttle valve.

The flow meter was developed with the aim of being able to measure the air mass admitted into the engine at each stroke (one stroke being equal to one-half revolution).

The flow meter body was constructed as a duct with a circular section. Two piezo-electrical microphones were mounted at fixed points along its axis. The entry duct to the flow meter was equipped with guide vanes to maintain laminar flow through the meter duct in order to obtain accuracy. The microphones were mounted on aerodynamically shaped supports developed to cause minimal disturbance in the flow field during measurement.

When excited by a step in the voltage supplied from the control circuit, the micro-phones emitted ultrasonic energy. The maximum amplitude of the resulting ultrasonic wave corresponded to the natural frequency of the microphones, which was about 300,000 cycles per second.

The moment the ultrasonic energy emission stopped, the microphones began to work as receivers for each other's waves.

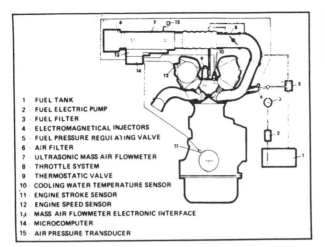

Marelli fuel-injection system
follows established lines except for the use of Marelli's
own exclusive development,
the ultrasonic flow meter.

1   FUEL TANK
2   FUEL ELECTRIC PUMP
3   FUEL FILTER
4   ELECTROMAGNETICAL INJECTORS
5   FUEL PRESSURE REGULATING VALVE
6   AIR FILTER
7   ULTRASONIC MASS AIR FLOWMETER
8   THROTTLE SYSTEM
9   THERMOSTATIC VALVE
10 COOLING WATER TEMPERATURE SENSOR
11 ENGINE STROKE SENSOR
12 ENGINE SPEED SENSOR
13 MASS AIR FLOWMETER ELECTRONIC INTERFACE
14 MICROCOMPUTER
15 AIR PRESSURE TRANSDUCER

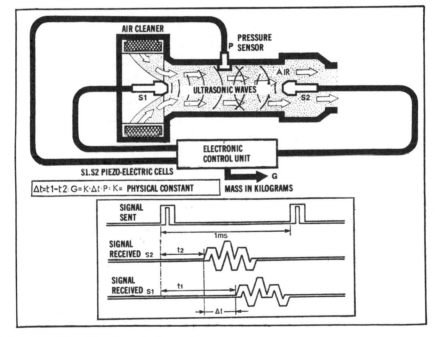

Schematic of Marelli's ultrasonic flow meter.

The two ultra-sound waves were emitted simultaneously, every millisecond. The relative delay between the moments of reception at opposite microphones (or, in other words, the difference in elapsed travel time of the two waves) provided information from which the air velocity could be calculated. By comparing the velocity signal with pressure and temperature readings, the wave could be made to express mass airflow.

Downstream from the meter duct was an acoustic filter, which had the task of protecting the high-frequency components from sound perturbations emanating from the engine.

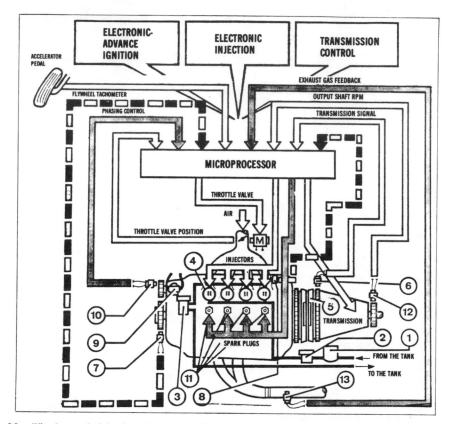

Marelli's electronic injection shares the electronic control unit with the electronic spark timing and gear-shifting. 1-Filter, 2-Pump, 3-Metering unit, 4-Injectors, 5-Clutch, 6-Transmission input speed sensor, 7-Engine rpm sensor, 8-Ignition distributor, 9-Coolant temperature sensor, 10-Camshaft position sensor, 11-Spark plugs, 12-Transmission output speed sensor, 13-Lambda-Sond sensor.

Flow meter tests proved that the unit was able to measure accurately (within a one-percent margin of error) flow masses throughout a range from 0.02 to 0.4 pound per second with velocities varying between 4.5 and 165 feet per second.

The error was attributed to boundary-layer conditions through the metered section of the duct. Ambient air temperature did not affect the flow pattern or the wave formation.

The control circuit picked up the time signals, which were first amplified, then filtered, and fed to logic-comparing circuits, from which they emerged transformed into a square electronic wave of specific length, proportional to mass airflow. In a second stage, this wave was processed further, compared with inputs from other transducers, resulting in an output that indicated air mass inflow during a single stroke. The necessary comparison data were supplied from the engine speed and stroke sensors.

The engine speed sensor was an induction coil which sensed the impulse frequency of magnetic pins attached to the flywheel.

The stroke sensor was an analog device which gave off a pulse every time it detected the passing of a reference mark on the camshaft. This mark signaled that the piston in cylinder number one was at top dead center, which gave the control unit a fix on the piston positions in all the other cylinders (in taking account of rpm signals).

Signals from the rpm sensor and stroke sensor were fed to a preprocessing block that converted them into basic logic data for synchronizing the complete system.

All the input data were translated into binary code before entering special registers where the microprocessor could read them.

The microprocessor was a digital computer capable of handling eight-bit words. It had two separate memories, a programmable read-only memory (PROM) and a random-access memory (RAM). The former contained the control program instructions, and the latter stored the input variables indicating the instantaneous engine-operating conditions.

The control-module output signals determined both the spark advance and the opening duration of the injector valves.

Fiat's experiments with this system, occurring in 1977, were aimed principally at reducing emission levels by maintaining a strictly constant air/fuel ratio (assisted by a Lambda-Sond oxygen sensor).

The ultrasonic airflow meter could also be combined with lean-burn engines aimed primarily at reduced fuel consumption. Further testing is needed to establish its potential in this regard.

Bosch/Pierburg
Electronic Carburetor

Just as a trend can be discerned by which fuel injection is moving closer to the carburetor, an opposite trend is also forming, by which the carburetor is evolving in the direction of fuel injection. Single-point injection systems are examples of the first trend. A carburetor with electronic fuel metering is the leading representative of the second.

Various carburetor companies began to look into the use of electronic fuel control for their carburetor systems as early as 1975, including Weber, Solex, Zenith, Stromberg, Carter and Holley.

Recently Pierburg formed a joint subsidiary with Bosch to develop and produce electronic control systems for conventional carburetors. Electronic controls have helped fuel-injection systems become more accurate, so as to reduce both emission level and fuel waste. They can do the same for the carburetor, say its proponents.

The Bosch/Pierburg system has provision for Lambda-Sond feedback, integral bypass systems, exhaust-gas recirculation, fuel shutoff on the overrun and spark timing from the same control unit.

The system is adaptable to all current types of carburetor (air-valve or constant-vacuum carburetors, as well as carburetors with variable throat geometry).

To control the actuators, a digital microprocessor capable of reading eight-bit words was selected. Sensors feed continuous data on engine speed, throttle position and coolant temperature.

The basic settings were developed to provide a lean mixture to give the best fuel economy with the least risk of abnormal combustion during steady-state operation with a hot engine. All other conditions are regarded merely as intermittent deviations from this type of operation, and their degree of fuel enrichment carefully worked out on the basis of supplementary control devices.

For warmup control, measurement devices are included to provide enrichment for cold-starting, after starting and after pick-up speed has been reached.

The exact systems differ to some extent according to type of carburetor used. In the constant-vacuum type, a fuel bypass system is provided, fully independent of the basic jet systems in the carburetor. The bypass system includes a vertical channel, a discharge nozzle and an electromagnetic valve.

Fuel flows from the discharge nozzle along the baffle close to the throttle plate. When the throttle is almost closed, the fuel passes directly through the hole in the throttle

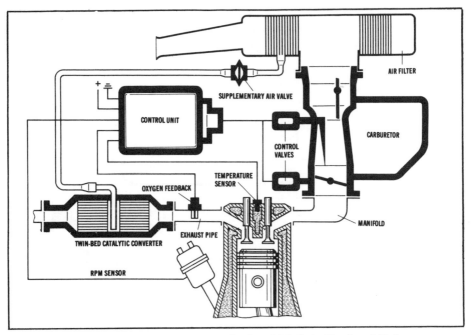

The electronically controlled carburetor system developed jointly by Bosch and Pierburg.

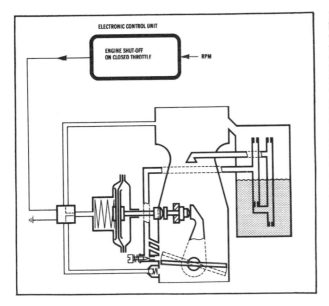

Fuel shutoff under engine-braking conditions is possible with the electronically controlled carburetor. The throttle valve is closed to a point beyond the normal idle position. An electromagnetic two-way or three-way valve takes control of the diaphragm that balances manifold vacuum against bypass air pressure and works the throttle-valve linkage as directed by orders from the electronic control unit. The programming of the control unit determines the fuel shutoff point, and when delivery is resumed.

plate, which is positioned in an area of high air velocity to provide fast and thorough atomization.

Increased airflow through the venturi is accompanied by higher vacuum, which increases fuel flow through the bypass system. A proportional control range can be achieved from idle to full load, due to the location of the discharge nozzle at a lower level than the fuel level in the float chamber.

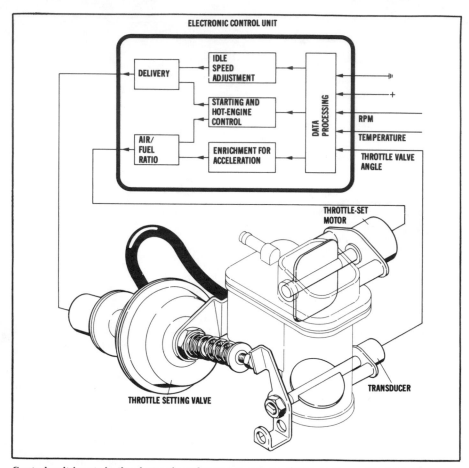

Control-unit inputs in the electronic carburetor system are similar to those of fuel-injection systems. It gets an rpm signal from the ignition distributor, engine temperature signals from a sensor in the coolant passage and throttle-valve position signals from a transducer on the throttle shaft. A choke plate operated by an electric motor, also computer-controlled, assures mixture enrichment for starting.

On variable-throat carburetors, a spring-loaded diaphragm is utilized to measure the vacuum in the venturi as a function of the mass airflow. A transducer relays this information to the control unit.

Fuel flows from the float chamber through an electromagnetic valve into a bypass channel, which delivers mixture to the discharge nozzle via an air bleed. A proportional control range over the total airflow range is assured by appropriate sizing and positioning of the fuel jets.

With all types of carburetor, an air bypass conducts air directly from the air cleaner to the intake manifold below the main throttle.

The bypass channel is equipped with its own throttle, which is automatically controlled by an electric motor. A potentiometer signals the main throttle angle to the control unit, which uses this information to adjust the bypass-throttle opening over the desired control range.

For cold-starting an automatic choke actuated by an electric motor sets itself according to the program in the control unit memory, as modified by a temperature signal.

Input data concerning engine speed, throttle position and temperature are compared with a time-since-startup factor on a continuous basis, gradually reducing the enrichment as soon as possible without risk of affecting the car's driveability.

With fast idle on cold-start, a stepper motor turns a spindle which acts on a control valve. The position of the spindle is transmitted by a pneumatic amplifier (needed to overcome the spring load of the throttle linkage) to the throttle plate.

Idle-speed control is assured by an aneroid capsule and electric pressure switch valves, which replace the conventional bypass mixture screw. The valves are operated by the electronic control unit.

Enrichment for acceleration is achieved by regulating the choke valve. On sudden opening of the throttle, it is closed to a predetermined angle and opens in a linear pattern as the car gains speed. Closing angle, dwell time and opening characteristics are adjustable.

Bosch and Pierburg engineers state that electronic controls on conventional carburetors offer important functional improvements as well as improvements in the stability of the setting and the durability of the system. The cost increase over conventional carburetors can be balanced by operational gains in fuel mileage combined with production-cost savings due to opportunities for simplifying the previously used emission controls.

Aftermarket Fuel Injection

By John Thawley

For the most part, the story of aftermarket fuel injection in America is curious and short-lived. The fuel injection companies were established by members of the Southern California hot rodding community in the fifties and early sixties. These were the days before large-capacity, high-flow four-barrel carbs and the hot rodders were on a constant search at this time for newer, better, lower-cost ways of feeding more and more mixture into their engines which were being raced on the dry lake beds and at drag strips. These were the days of up to eight single-throat carbs mounted on a V-8 engine. Six-carb manifolds for V-8's were very common.

Stu Hilborn is generally credited with being the pioneer in developing hot rod fuel injection. The company survives today under the name of Fuel Injection Engineering. A variety of Chevrolet, Ford, Chrysler, Volkswagen and motorcycle systems are available.

The Hilborn injector system is a basically simple one. It consists primarily of a short, cylindrical throttle body containing the injector nozzle and butterfly, a needle-type metering valve linked to and operated in conjunction with the butterfly, a bypass or bleed jet which controls the overall amount of fuel supplied to the mixture valve, and the engine-driven pump which supplies the system with a continuous flow of fuel under pressure.

In operation, the pump delivers fuel to the metering valve and injector nozzle at a rate of flow proportioned to the engine speed. The metering valve also serves to adjust the rate of fuel flow. Between the pump and the metering valve, the bypass jet is located. This is simply a calibrated orifice which drains off a predetermined amount of the pressurized fuel and returns it to the fuel tank. It provides overall mixture control by changing the size of the orifice. For a leaner overall mixture a larger opening is used; a smaller one will supply a richer mixture.

The Hilborn system will not supply uniform mixture control over the entire speed range of an engine, but it is efficient over a relatively narrow speed range, particularly at high rpm. Today Hilborn injectors are found primarily on sprint cars and on dry lakes and Bonneville race cars.

This is a Hilborn unit with velocity stacks removed. Note the relatively massive throttle linkage which is adjustable at several points.

This is a Crower fuel-injection system for a small-block Chevy. A similar unit is available for a big-block Chevy. The aluminum port stacks are available in a variety of lengths to help tune the unit to a particular application.

The underside of an Enderle 'butterfly' housing reveals the eight injector nozzles which squirt fuel into a GMC-type blower. This is for drag race application only.

The design inspiration behind the development of the Enderle injector system came from Kent Enderle and Pete Jackson. The Enderle design has been very successful in improving mixture control at the upper end of the engine's speed range. This is extremely critical in blown fuel applications in drag racing and this is where the Enderle system continues to reign supreme. The butterfly housing sits atop a GMC-type blower which in turn is bolted to the engine cylinder heads. Other components of the Enderle system are the injector nozzles installed in the throttle body housing, a fuel-control-valve assembly linked to the throttle plate shafts as an integral part of the assembly, a fuel shutoff valve, a filter and a belt-driven impeller-type fuel pump. In operation, the pump pulls the fuel from the tank, through the shutoff valve and filter and forces it into the fuel control valve. The volume and pressure of the fuel increases proportionately with engine speed. The fuel may follow one of several paths at the control valve. If the throttle plates are closed to idle position, a passage in the valve's rotating piston aligns with the inlet port and tank return port so the unused fuel may return to the tank. Some fuel is allowed by the idle system to reach the injector nozzles in the throttle body so the engine can sustain low-speed operation.

As the throttle plates and control valve piston are moved to the fully open position, the tank return port is closed and a second passage opens leading to the injector nozzles. Some slight metering of the fuel takes place as the passage between pump and injector nozzles is opened, but this is rather incidental to the system. There is no overt pretense or provision for fuel metering under part-throttle conditions. The system is primarily intended to operate either all-on or all-off. Like the Hilborn system, a bypass jet is used to control mixture richness, but if the mixture is set either rich or lean, it will remain so over the range from idle to top rpm. The Enderle system is the most popular fuel injection for all supercharged applications in racing—tractor pulls, boat and car drag racing in all classes.

The Scott injector is one of several aftermarket fuel-injection systems that has passed from the scene. The Scott was interesting in that it utilized a centrifugal metering pump rather than a positive displacement pump. The Scott centrifugal pump squared the pressure as the engine rpm doubled. This compensated for the pressure drop caused

There are several styles of air cleaners for injector systems. Both of these are built by K&N. Both of these injected small-block Chevys are in sprint cars which race only on dirt.

This quick-disconnect main bypass available from Kinsler allows very quick jet changes without the use of any wrenches.

This is an adjustable-diaphragm high-speed bypass valve which can be used to replace the jet can-type bypass. Turning the adjustment screw in raises the pressure (rpm) at which the valve opens. This makes the system run richer.

by the nozzles and gave a consistent air/fuel ratio throughout the entire rpm range. The Scott system was sold to the Mickey Thompson Equipment Company in the mid-sixties and was marketed for some time under the name of M/T Scott Superslot before disappearing from the marketplace.

The Orner system was around briefly in the early sixties and then faded quietly from the ranks of racers. The few Orner systems built were all for supercharged applications except for some experimental models built for normally aspirated use. None of the latter units were sold. The Orner system differed from all others in that fuel metering was accomplished by a manifold pressure-operated fuel-injection valve which regulated fuel flow in accordance with density changes within the manifold. The valve was mounted on the injector housing and contained a rubber diaphragm which actuated a spring-loaded pintle-type needle valve. When manifold pressure went above atmospheric, the diaphragm moved the needle valve toward the open position to permit a greater fuel flow. When pressure in the manifold dropped, the diaphragm/needle arrangement restricted the fuel flow. Like the Hilborn and Enderle systems a bypass jet was used to control overall mixture conditions.

The Algon system rests peacefully with the Scott and Orner systems. In the early sixties Algon injectors could be purchased for any one of several domestic V-8 engines being built at that time. They were available for normally aspirated and for blown applications. The air-control system was similar to the Hilborn, but the fuel system differed in some respects. Fuel pressure was supplied by a constant-pressure aircraft pump that

The Edelbrock EFI (electronic fuel injection) was produced for a short period of time for a limited number of small-block Chevy applications. Throttle response and fuel economy were excellent.

could be adjusted to deliver a pressure within a range of five to fifteen pounds. This is in contrast to the thirty pounds of pressure found in Hilborn systems on normally aspirated engines and up to sixty pounds in blown applications.

A tapered needle within the Algon's regulating valve controlled the flow of fuel by its moving back and forth in an orifice by linkage connected to the throttle-valve mechanism. Additional fuel metering could be done at the discharge nozzles with tapered needles that were moved in and out of orifices by rotating threaded caps to which they were connected.

Tecalemit, McKay and Jackson were all fairly straightforward variations of the Hilborn design. None of these systems are presently with us. One Hilborn copy which is alive and well is the Crower—currently available for large- and small-block Chevys in addition to Volkswagens.

Currently, Edelbrock is marketing an electronic fuel-injection system for a select number of applications of the small-block Chevy. Although the system has been dropped from its current catalog some words concerning the Edelbrock EFI are in order. In the first place it is the most sophisticated piece of performance equipment ever to be produced by an aftermarket manufacturing firm. The injector plumbing system is controlled by a 'black box.' A mini-computer and an 'injector drive board' provide injector-on and injector-off data to the electrically operated injectors. A pressure transducer produces intake manifold vacuum data which is part of the computed fuel signal data. A number of other engine-operating variables such as inlet air temperature, engine rpm, ignition advance and coolant temperature are constantly monitored and the information is fed into the mini-computer, which in turn changes the signal going to the injector nozzle and alters the length of the injector 'on-time.' This constant altering goes on at the rate of 1,000 times per second! Additionally the fuel-injection system is tied directly to the ignition distributor for very precise control over ignition timing as related to all of the operating variables of the engine. The end result of all of this electronic sophistication is greatly improved fuel

This jet selector valve allows the driver to richen or lean the mixture during the race. On this particular valve there are eight different selections.

economy and throttle response. However, expense and vehicle application limitations have greatly limited the sales of the system.

Oddly, the premier aftermarket fuel-injection firm in this country today does not manufacture a complete fuel-injection system for any engine. Kinsler Fuel Injection services and sells parts for Hilborn, Crower, Enderle and even Lucas. Jim Kinsler is recognized as the leading authority in this country on domestically produced fuel injection and regularly stocks the largest assortment of parts for all systems—including those long since out of production. This firm routinely builds and services custom fuel-injection systems for one-off road race cars, turbocharged engines of all types and is keenly attuned to what Detroit is looking at in the way of OEM fuel-injection systems in the future.

Crower
3333 Main St.
Chula Vista, CA 92011

Edelbrock
411 Coral Circle
El Segundo, CA 90245

Enderle Fuel Injection
6840 E. DeBie Drive
Paramount, CA 90723

Fuel Injection Engineering
25891 Crown Valley Parkway
South Laguna, CA 92677

Kinsler Fuel Injection
1834 Thunderbird
Troy, MI 48084

Appendices

1981-Model Cars Using L-Jetronic

Make	Model	No. cyl.	Bore x Stroke	Cu. In.	Hp @ rpm
BMW	528i	6	3.39x3.15	170	184@5800
BMW	535i	6	3.68x3.31	210.7	218@5200
BMW	628 CSi	6	3.39x3.15	170	184@5800
BMW	633 CSi	6	3.50x3.39	196	197@5500
BMW	728i	6	3.39x3.15	170	184@5800
Citroen	CX 2400	4	3.68x3.37	143	128@4800
Citroen	CX GTi	4	3.68x3.37	143	128@4800
Citroen	CX Pallas	4	3.68x3.37	143	128@4800
Citroen	CX Prestige	4	3.68x3.37	143	128@4800
Daimler	Sovereign	6	3.63x4.17	258.4	208@5000
Fiat	Argenta	4	3.31x3.54	121.7	122@5300
Fiat	124 Spider	4	3.31x3.54	121.7	122@5300
Jaguar	XJ-6 4.2	6	3.63x4.17	258.4	208@5000
Isuzu	Gemini	4	3.31x3.23	111	130@6400
Isuzu	Florian	4	3.43x3.23	119	135@6200
Lancia	Trevi	4	3.31x3.54	121.7	122@5500
Lancia	Gamma 2500 IE	4	4.02x2.99	151.6	140@5400
Mazda	Luce SG	4	3.15x3.86	120	120@5500
Nissan	Violet	4	3.27x2.90	97	105@6000
Nissan	Violet	4	3.35x3.07	108	115@6000
Nissan	Auster	4	3.27x2.90	97	105@6000
Nissan	Auster	4	3.35x3.07	108	115@6000
Nissan	Stanza	4	3.27x2.90	97	105@6000
Nissan	Stanza	4	3.35x3.07	108	115@6000
Nissan	Silvia	4	3.35x3.07	108	115@6000
Nissan	Silvia	4	3.35x3.39	119	120@5600
Nissan	Gazelle	4	3.35x3.07	108	115@6000

Nissan	Gazelle	4	3.35x3.39	119	120@5600
Nissan	Bluebird	4	3.35x3.07	108	115@6000
Nissan	Bluebird	4	3.35x3.39	119	120@5600
Nissan	Bluebird	4	3.35x3.07	108	135@6000
Nissan	Skyline	4	3.35x3.07	108	115@6000
Nissan	Skyline	6	3.07x2.74	122	130@6000
Nissan	Skyline	6	3.07x2.74	122	145@5600
Nissan	Cedric	6	3.07x2.74	122	130@6000
Nissan	Cedric	6	3.07x2.74	122	145@5600
Nissan	Cedric	6	3.39x3.11	168	145@5200
Nissan	Gloria	6	3.07x2.74	122	130@6000
Nissan	Gloria	6	3.07x2.74	122	145@5600
Nissan	Gloria	6	3.39x3.11	168	145@5200
Nissan	Laurel	6	3.07x2.74	122	125@6000
Nissan	Laurel	6	3.07x2.74	122	145@5600
Nissan	Laurel	6	3.39x3.11	168	155@5200
Nissan	President	V-8	3.62x3.27	269	200@4800
Nissan	Fairlady Z	6	3.07x2.74	122	130@6000
Nissan	Fairlady Z	6	3.07x2.74	122	145@5600
Opel	Ascona SR 2.0 E	4	3.74x2.75	121	110@5400
Opel	Manta GT E	4	3.74x2.75	121	110@5400
Opel	Ascona 400	4	3.74x2.75	121	144@5200
Opel	Rekord E	4	3.74x2.75	121	110@5400
Opel	Senator E	6	3.74x2.75	181	180@5800
Opel	Monza E	6	3.74x2.75	181	180@5800
Toyota	Corolla	4	3.35x2.76	97	115@6000
Toyota	Sprinter	4	3.35x2.76	97	115@6000
Toyota	Carina	4	3.35x2.76	97	115@6000
Toyota	Carina	4	3.48x3.15	120	135@5800
Toyota	Celica	4	3.35x2.76	97	115@6000
Toyota	Celica	4	3.48x3.15	120	135@5800
Toyota	Celica	6	2.95x2.95	121	125@6000
Toyota	Celica	4	3.27x3.35	169	145@5000
Toyota	Corona	4	3.48x3.15	120	135@5800
Toyota	Crown	6	2.95x2.95	121	125@6000
Toyota	Crown	6	2.95x2.95	121	145@5600
Toyota	Crown	6	3.27x3.35	169	145@5000
Toyota	Chaser	4	3.48x3.15	120	135@5800
Toyota	Chaser	6	2.95x2.95	121	125@5400
Toyota	Chaser	6	3.27x3.35	169	145@6000
Toyota	Cresta	6	2.95x2.95	121	125@6000
Toyota	Century	V-8	3.27x3.07	206	180@5200

1981-Model Cars Equipped With Motronic As Standard

Make	Model	No. cyl.	Bore x Stroke	Cu. In.	Hp @rpm
BMW	732i	6	3.50x3.39	196	197@5500
BMW	735i	6	3.68x3.31	211	218@5200
BMW	745i	6	3.50x3.39	196	252@5200

1981-Model Cars Equipped With Kugelfischer Injection

Make	Model	No. cyl.	Bore x Stroke	Cu. In.	Hp @ rpm
BMW	M-1	6	3.68x3.31	211	277@6500
Peugeot	504 Coupe	4	3.46x3.19	120	101@5200
Peugeot	504 Cabriolet	4	3.46x3.19	120	101@5200

1981-Model Cars Using K-Jetronic

Make	Model	No. cyl.	Bore x Stroke	Cu. In.	Hp @ rpm
Audi	80 GLE	4	3.13x3.15	97	110@6100
Audi	100 5 E	5	3.13x3.40	131	136@5700
Audi	200	5	3.13x3.40	131	136@5700
Audi	200 5T	5	3.13x3.40	131	170@5300
Audi	Quattro	5	3.13x3.40	131	200@5500
Bentley	Mulsanne	V-8	4.10x3.90	412	Not available
BMW	318i	4	3.50x2.80	108	105@5800
BMW	323i	6	3.15x3.02	141	143@5800
De Lorean	DMC-12	V-6	3.46x2.87	163	137@5500
Ferrari	308 GTB i	V-8	3.19x2.80	178.5	214@6600
Ferrari	Mondial 8	V-8	3.19x2.80	178.5	214@6600
Ferrari	400 Automatic	V-12	3.19x3.07	294	310@6400
Ford	Granada S	V-6	3.66x2.70	170	160@5700
Ford	Capri 2.8 i	V-6	3.66x2.70	170	160@5700
Mercedes-Benz	230 E	4	3.76x3.16	140	136@5100
Mercedes-Benz	280 E	6	3.39x3.10	168	185@5800
Mercedes-Benz	280 SE	6	3.39x3.10	168	185@5800
Mercedes-Benz	380 SE	V-8	3.62x2.83	233	218@5500
Mercedes-Benz	500 SE	V-8	3.82x3.35	303.5	240@4750
Peugeot	504 Coupe V-Six	V-6	3.46x2.87	163	144@5500
Peugeot	505 STI	4	3.46x3.23	122	110@5250
Peugeot	604 STI	V-6	3.46x2.87	163	144@5500
Porsche	924	4	3.41x3.32	121	125@5800
Porsche	924 Turbo	4	3.41x3.32	121	177@5500
Porsche	924 Carrera GT	4	3.41x3.32	121	210@6000
Porsche	924 Carrera GTS	4	3.41x3.32	121	245@6250
Porsche	911 SC	6	3.74x2.77	183	204@5900
Porsche	911 Turbo	6	3.82x2.93	201.3	300@5500
Porsche	928	V-8	3.74x3.11	273	240@5250
Porsche	928 S	V-8	3.82x3.11	284.6	300@5900
Renault	5 Turbo	4	2.99x3.03	85.2	160@6000
Rolls-Royce	Silver Spirit	V-8	4.10x3.90	412	Not available
Saab	900 E	4	3.54x3.07	121	118@5500
Saab	900 EMS	4	3.54x3.07	121	118@5500
Saab	900 Turbo	4	3.54x3.07	121	145@5000
Volkswagen	Golf GTI	4	3.13x3.15	97	110@6100
Volkswagen	Scirocco GT	4	3.13x3.15	97	110@6100
Volvo	242 GLE	4	3.62x3.15	130	140@5500
Volvo	244 Turbo	4	3.62x3.15	130	155@5500
Volvo	264 GLE	V-6	3.58x2.87	174	155@5500

Index

More Great Reading

The Cobra Story. Autobiography of Carroll Shelby and Cobra production & racing history through 1965. 272 pages, 60 photos.

American Car Spotter's Guide 1966-1980. Giant pictorial source with over 3,600 illustrations. 432 pages, softbound.

Ferraris For The Road. Provides lavish pictorial coverage of Ferrari's production models. In the Survivors Series. 126 pages, 269 photos, many in color.

Porsches For The Road. Includes photo essays on 12 models. In the Survivors Series. 128 pages, 250 illus., 125 in color.

Shelby's Wildlife: The Cobras and Mustangs. Complete, exciting story of the 260, 289, 427 and Daytona Cobras plus Shelby Mustangs. 224 pages, nearly 200 photos.

Chevy Super Sports 1961-1976. Exciting story of these hot cars with complete specs and data. 178 pages, 198 illus., softbound.

The Art & Science of Grand Prix Driving. Incisive look at F1 cars and techniques by world driving champion Niki Lauda. 245 pages, 158 photos, 23 in color.

Muscle Car Mania: An Advertising Collection. 250 of those great promotional ads for 1964 through 1974 muscle cars. Softbound, 176 pages.

The Production Figure Book for U.S. Cars. Reflects the relative rarity of various makes, models, body styles, etc. Softbound, 180 pages.

Imported Car Spotter's Guide. Over 2,000 illustrations from 83 manufacturers in 11 countries, through 1979. 359 pages, softbound.

Lincoln and Continental: The Postwar Years. Interesting historical information through 1980. 152 pages, 223 illustrations.

Pontiac: The Postwar Years. One of America's most exciting makes is covered in this factual story. 205 pages, 222 photos.

Illustrated Ferrari Buyer's Guide. Features all street production models 1954 through 1980. 169 pages, over 230 photos, softbound.

Auto Restoration From Junker To Jewel. Illustrated guide to restoring old cars. 232 pages, 289 illustrations, softbound.

Oldsmobile: The Postwar Years. 280 fine illustrations help tell this exciting story through 1980. 152 pages.